Guide to
ANALYSIS OF DNA
MICROARRAY DATA

Guide to
ANALYSIS OF DNA
MICROARRAY DATA

Second Edition

Steen Knudsen

Center for Biological Sequence Analysis
BioCentrum-DTU

Technical University of Denmark

WILEY-LISS

A JOHN WILEY & SONS, INC., PUBLICATION

Library of Congress Cataloging-in-Publication Data is available.

ISBN 0-471-65604-6

Printed in the United States of America.

10 9 8 7 6 5 4 3 2 1

To Linnea

Contents

Preface

I am often asked, "Do you have a good text I can read on analysis of DNA array data?" This is an attempt at providing such a text for students and scientists alike who venture into the field of DNA array data analysis for the first time. The book is written for biologists and medical researchers without special training in data analysis and statistics. Mathematical stringency is sacrificed for intuitive and visual introduction of concepts. Methods are introduced by simple examples and citations of relevant literature. Practical computer solutions to common analysis problems are suggested, with an emphasis on software developed at and made freely available by my own lab. The text emphasizes gene expression analysis.

This text takes over where the DNA array equipment leaves you: with a file containing an image of the microarray. If the equipment has already performed an analysis of the image, you are left with a file of signal intensities. The information in that file will prompt questions such as: How is it scaled? What is the error in the data? When can I say that a certain gene is up-regulated? What do I do with the thousands of genes that show some regulation? How much information can I get out of my data? This text will attempt to answer those questions and others that will come into mind as you delve further into the data.

Since the appearance of the first edition, the field has virtually exploded, with thousands of papers published on DNA microarrays and data analysis. A new generation of microarray equipment, allowing *in situ* synthesis of chips, has appeared. New public software packages have appeared, and improved methods for data analysis have been published. The second edition includes all these new and recent developments and also contains new chapters on image analysis, experiment design, interpretation of results, oligonucleotide probe design, data integration, and systems biology. The second edition aims to be the most comprehensive and up-to-date book available on DNA microarrays.

Each chapter has a section on Further Reading, which categorizes key literature by topic.

A web companion site[1] is available with copy-paste code examples from the book, errata, experimental protocols, and more.

STEEN KNUDSEN

Lyngby, Denmark
December 2003

---

[1] http://www.cbs.dtu.dk/steen/book.html

Acknowledgments

Christopher Workman, Laurent Gautier, and Henrik Bjørn Nielsen inspired me for many aspects of this book and also implemented many methods used in the book.

I thank Yves Moreau for helpful suggestions on the manuscript.

I thank my collaborators Claus Nielsen, Kenneth Thirstrup, Torben Ørntoft, Friedrik Wikman, Thomas Thykjaer, Mogens Kruhøffer, Karin Demtröder, Hans Wolf, Lars Dyrskjøt Andersen, Casper Møller Frederiksen, Jeppe Spicker, Lars Juhl Jensen, Carsten Friis, Hanne Jarmer, Hans-Henrik Saxild, Randy Berka, Matthew Piper, Steen Westergaard, Christoffer Bro, Thomas Jensen, and Kristine Dahlin for allowing me to use examples generated from our collaborative research.

I am grateful to Center Director Søren Brunak for creating the environment, and to the Danish National Research Foundation, the Danish Biotechnology Instrument Center, and Novozymes A/S for funding the research that made this book possible.

S. K.

1

Introduction to DNA Microarray Technology

1.1 HYBRIDIZATION

The fundamental basis of DNA microarrays is the process of *hybridization*. Two DNA strands hybridize if they are complementary to each other. Complementarity reflects the Watson-Crick rule that adenine (A) binds to thymine (T) and cytosine (C) binds to guanine (G). One or both strands of the DNA hybrid can be replaced by RNA and hybridization will still occur as long as there is complementarity.

Hybridization has for decades been used in molecular biology as the basis for such techniques as Southern blotting and Northern blotting. In Southern blotting, a small string of DNA, an *oligonucleotide*, is used to hybridize to complementary fragments of DNA that have been separated according to size in a gel electrophoresis. If the oligonucleotide is radioactively labeled, the hybridization can be visualized on a photographic film that is sensitive to radiation. In Northern blotting, a radio-labeled oligonucleotide is used to hybridize to messenger RNA that has been run through a gel. If the oligo is specific to a single messenger RNA, then it will bind to the location (*band*) of that messenger in the gel. The amount of radiation captured on a photographic film depends to some extent on the amount of radio-labeled probe present in the band, which again depends on the amount of messenger. So this method is a semiquantitative detection of individual messengers.

DNA arrays are a massively parallel version of Northern and Southern blotting. Instead of distributing the oligonucleotide probes over a gel containing samples of RNA or DNA, the oligonucleotide probes are attached

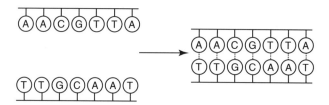

Fig. 1.1 Hybridization of two DNA molecules. Dotted line: hydrogen bonds.

to a surface. Different probes can be attached within micrometers of each other, so it is possible to place many of them on a small surface of one square centimeter, forming a DNA array. The sample is labeled fluorescently and added to the array. After washing away excess unhybridized material, the hybridized material is excited by a laser and is detected by a light scanner that scans the surface of the chip. Because you know the location of each oligonucleotide probe, you can quantify the amount of sample hybridized to it from the image generated by the scan.

There is some contention in the literature on the use of the word "probe" in relation to microarrays. Throughout this book the word "probe" will be used to refer to what is attached to the microarray surface. And the word "target" will be used to refer to what is hybridized to the probes.

Where before it was possible to run a couple of Northern blots or a couple of Southern blots in a day, it is now possible with DNA arrays to run hybridizations for tens of thousands of probes. This has in some sense revolutionized molecular biology and medicine. Instead of studying one gene and one messenger at a time, experimentalists are now studying many genes and many messengers at the same time. In fact, DNA arrays are often used to study *all* known messengers of an organism. This has opened the possibility of an entirely new, systemic view of how cells react in response to certain stimuli. It is also an entirely new way to study human disease by viewing how it affects the expression of all genes inside the cell. Figure 1.2 illustrates the revolution of DNA arrays in biology and medicine by the number of papers published on the topic.

1.2 GOLD RUSH?

The explosion in interest in DNA microarrays has almost been like a gold rush. Is there really that much gold to be found with this new technology? I am afraid that, in the short term, there will be some disappointments. Yes, you can learn about the gene expression in your organism or disease of interest,

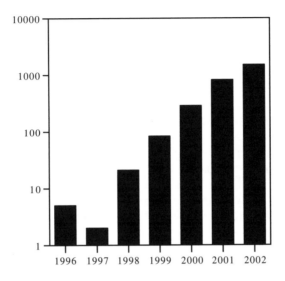

Fig. 1.2   The number of papers published per year referring to DNA microarrays.

but does that make you wiser? Typically, the wealth of data generated results in more questions than answers. There is one exception to this, and that is where DNA arrays have been used for diagnostics and prognostics. Here, DNA arrays have shown promising results in almost all the fields where they have been applied. This is where I think that the greatest short-term success of DNA microarray technology lies.

On a longer time scale molecular biology will benefit tremendously from the systemic approach offered by DNA microarrays and other massively parallel approaches. Many important discoveries lie in the interpretation of microarray data – more so from large compilations of experiments and large-scale experiments than from small experiments with just a few arrays.

1.3   THE TECHNOLOGY BEHIND DNA MICROARRAYS

When DNA microarrays are used for measuring the concentration of messenger RNA in living cells, a *probe* of one DNA strand that matches a particular messenger RNA in the cell is used. The concentration of a particular messenger is a result of *expression* of its corresponding gene, so this application is often referred to as *expression analysis*. When different probes matching all messenger RNAs in a cell are used, a snapshot of the total messenger RNA pool of a living cell or tissue can be obtained. This is often referred to as an *expression profile* because it reflects the expression of every single measured gene at that particular moment. Expression profile is also sometimes used to describe the expression of a single gene over a number of conditions.

Expression analysis can also be performed by a method called *serial analysis of gene expression* (SAGE). Instead of using microarrays, SAGE relies on traditional DNA sequencing to identify and enumerate the messenger RNAs in a cell (see Section 1.3.6).

Another traditional application of DNA microarrays is to detect mutation in specific genes. The massively parallel nature of DNA microarrays allows the simultaneous screening of many, if not all, possible mutations within a single gene. This is referred to as *genotyping* (Chapter 12).

The treatment of array data does not depend so much on the technology used to gather the data as it depends on the application in question. Genotyping and expression analysis are two completely different applications, and they will be treated separately in this text. Most of the information will address analysis of expression data, and a separate chapter will address genotyping chips.

For expression analysis the field has been dominated in the past by two major technologies. The Affymetrix, Inc. GeneChip system uses prefabricated oligonucleotide chips (Figures 1.3 and 1.4). Custom-made chips use a robot to spot cDNA, oligonucleotides, or PCR products on a glass slide or membrane(Figure 1.5).

More recently, several new technologies have entered the market. In the following, several of the major technology platforms for gene expression analysis will be described.

1.3.1 Affymetrix GeneChip Technology

Affymetrix uses equipment similar to that which is used for making silicon chips for computers, and thus allows mass production of very large chips at reasonable cost. Where computer chips are made by creating masks that control a photolithographic process for removal or deposition of silicon material on the chip surface, Affymetrix uses masks to control synthesis of oligonucleotides on the surface of a chip. The standard phosphoramidite method for synthesis of oligonucleotides has been modified to allow light control of the individual steps. The masks control the synthesis of several hundred thousand squares, each containing many copies of an oligo. So the result is several hundred thousand different oligos, each of them present in millions of copies.

That large number of oligos, up to 25 nucleotides long, has turned out to be very useful as an experimental tool to replace all experimental detection procedures that in the past relied on using oligonuclotides: Southern, Northern, and dot blotting as well as sequence specific probing and mutation detection.

For expression analysis, up to 40 oligos are used for the detection of each gene. Affymetrix has chosen a region of each gene that (presumably) has the

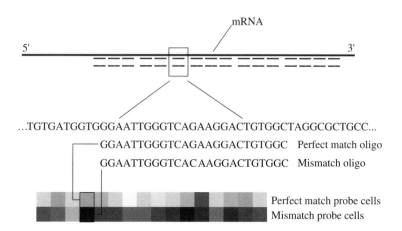

Fig. 1.3   The Affymetrix GeneChip technology. The presence of messenger RNA is detected by a series of probe pairs that differ in only one nucleotide. Hybridization of fluorescent messenger RNA to these probe pairs on the chip is detected by laser scanning of the chip surface. (Figure by Christoffer Bro.)

least similarity to other genes. From this region 11 to 20 oligos are chosen as perfect matches (PM) (i.e., perfectly complementary to the mRNA of that gene). In addition, they have generated 11 to 20 mismatch oligos (MM), which are identical to the PM oligos except for the central position 13, where one nucleotide has been changed to its complementary nucleotide. Affymetrix claims that the MM oligos will be able to detect nonspecific and background hybridization, which is important for quantifying weakly expressed mRNAs. However, for weakly expressed mRNAs where the signal-to-noise ratio is smallest, subtracting mismatch from perfect match adds considerably to the noise in the data (Schadt et al., 2000). That is because subtracting one noisy signal from another noisy signal yields a third signal with even more noise.

The hybridization of each oligo to its target depends on its sequence. All 11 to 20 PM oligos for each gene have a different sequence, so the hybridization will not be uniform. That is of limited consequence as long as we wish to detect only *changes* in mRNA concentration between experiments. How such a change is calculated from the intensities of the 22 to 40 probes for each gene will be covered in Section 4.3.

To detect hybridization of a target mRNA by a probe on the chip, we need to label the target mRNA with a fluorochrome. As shown in Figure 1.4, the steps from cell to chip usually are as follows:

- Extract total RNA from cell (usually using TRIzol from Invitrogen or RNeasy from QIAGEN).

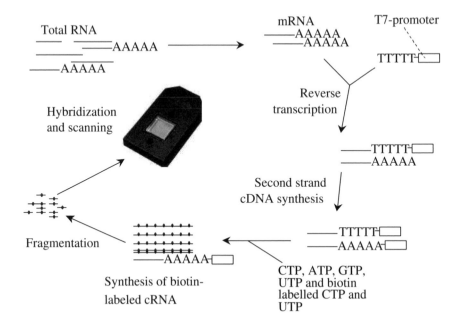

Fig. 1.4 Preparation of sample for GeneChip arrays. Messenger RNA is extracted from the cell and converted to cDNA. It then undergoes an amplification and labeling step before fragmentation and hybridization to 25-mer oligos on the surface of the chip. After washing of unhybridized material, the chip is scanned in a confocal laser scanner and the image analyzed by computer. (Figure by Christoffer Bro.)

- Separate mRNA from other RNA using poly-T column (optional).

- Convert mRNA to cDNA using reverse transcriptase and a poly-T primer.

- Amplify resulting cDNA using T7 RNA polymerase in the presence of biotin-UTP and biotin-CTP, so each cDNA will yield 50 to 100 copies of biotin-labeled cRNA.

- Incubate cRNA at 94 degrees in fragmentation buffer to produce cRNA fragments of length 35 to 200 nucleotides.

- Hybridize to chip and wash away non-hybridized material.

- Stain hybridized biotin-labeled cRNA with Streptavidin-Phycoerythrin and wash.

- Scan chip in confocal laser scanner (optional).

Table 1.1   Performance of the Affymetrix GeneChip technology. Numbers refer to chips in routine use and the current limit of the technology (Lipshutz et al., 1999; Baugh et al., 2001).

	Routine use	Current limit
Starting material	5 μg total RNA	2 ng total RNA
Detection specificity	$1 : 10^5$	$1 : 10^6$
Difference detection	twofold changes	10% changes
Discrimination of related genes	70–80% identity	93% identity
Dynamic range (linear detection)	3 orders of magn.	4 orders of magn.
Probe pairs per gene	20	4
Number of genes per array	12,000	40,000

- Amplify the signal on the chip with goat IgG and biotinylated antibody.

- Scan chip in scanner.

Usually, 5 to 10 μg of total RNA are required for the procedure. But new improvements to the cDNA synthesis protocols reduce the required amount to 100 ng. If two cycles of cDNA synthesis and cRNA synthesis are performed, the detection limit can be reduced to 2 ng of total RNA (Baugh et al., 2001). MessageAmp kits from Ambion allow up to 1000 times amplification in a single round of T7 polymerase amplification. The current performance of the Affymetrix GeneChip technology is summarized in Table 1.1.

1.3.2   Spotted Arrays

In another major technology, spotted arrays, a robot spotter is used to move small quantities of probe in solution from a microtiter plate to the surface of a glass plate. The probe can consist of cDNA, PCR product, or oligonucleotides. Each probe is complementary to a unique gene. Probes can be fixed to the surface in a number of ways. The classical way is by non-specific binding to polylysine-coated slides. The steps involved in making the slides can be summarized as follows (Figure 1.5):

- Coat glass slides with polylysine.

- Prepare probes in microtiter plates.

- Use robot to spot probes on glass slides.

- Block remaining exposed amines of polylysine with succinic anhydride.

- Denature DNA (if double-stranded) by heat.

The steps involved in preparation of sample and hybridizing to the array can be summarized as follows (Figure 1.5):

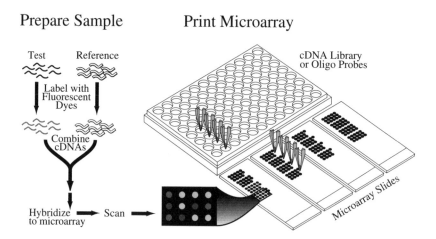

Prepare Sample   **Print Microarray**

Fig. 1.5   The spotted array technology. A robot is used to transfer probes in solution from a microtiter plate to a glass slide where they are dried. Extracted mRNA from cells is converted to cDNA and labeled fluorescently. Reference sample is labeled red and test sample is labeled green. After mixing, they are hybridized to the probes on the glass slide. After washing away unhybridized material, the chip is scanned with a confocal laser and the image analyzed by computer. (See color plate.)

- Extract total RNA from cells.
- Optional: isolate mRNA by polyA tail.
- Convert to cDNA in the presence of Amino-allyl-dUTP (AA-dUTP).
- Label with Cy3 or Cy5 fluorescent dye linking to AA-dUTP.
- Hybridize labeled mRNA to glass slides.
- Wash unhybridized material away.
- Scan slide and analyze image (see example image in Figure 1.6).

The advantage compared to Affymetrix GeneChips is that you can design any probe for spotting on the array. The disadvantage is that spotting will not be nearly as uniform as the *in situ* synthesized Affymetrix chips and that the cost of oligos, for chips containing thousands of probes, becomes high. From a data analysis point of view, the main difference is that in the cDNA array usually the sample and the control are hybridized to the same chip using different fluorochromes, whereas the Affymetrix chip can handle only one fluorochrome so two chips are required to compare a sample and a control.

Table 1.2 shows the current performance of the spotted array technology

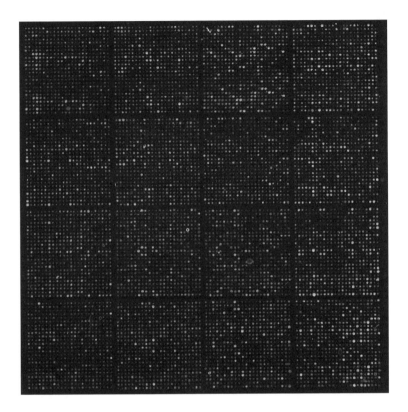

Fig. 1.6 Spotted array containing more than 9000 features. Probes against each predicted open reading frame in *Bacillus subtilis* are spotted twice on the slide. Image shows color overlay after hybridization of sample and control and scanning. (See color plate. Picture by Hanne Jarmer.)

1.3.3   Digital Micromirror Arrays

In 1999, Singh-Gasson et al. published a paper in *Nature Biotechnology* showing the feasibility of using digital micromirror arrays to control light-directed synthesis of oligonucleotide arrays. Two commercial companies were formed based on this technology. NimbleGen (www.nimblegen.com) synthesizes DNA arrays using digital micromirrors and sells the manufactured arrays to the customer. Febit (www.febit.de) makes an instrument, the

Table 1.2 Performance of the spotted array technology (Schena, 2000).

	Routine use
Starting material	10–20 μg total RNA
Dynamic range (linear detection)	3 orders of magnitude
Number of probes per gene	1
Number of genes or ESTs per array	$\approx$10,000

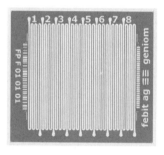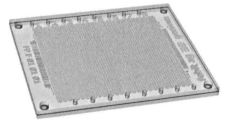

Fig. 1.7 The Febit DNA processor with microchannel structure. Left a two-dimensional view of the microchannels. Right a three-dimensional view of the microchannels including inlet and outlet of each channel. Copyright Febit AG. Used with permission.

Geniom One, which allows the customer to control digital micromirror synthesis in his own lab (Baum et al., 2003). The design of the microarray is uploaded by computer and the synthesis takes about 12 hours in a DNA processor (Figure 1.7 and 1.8). Then the fluorescently labeled sample is added, and after hybridization and washing the fluorescence is read by a CCD camera and the resulting image returned to the computer. All steps are integrated into a single instrument.

1.3.4 Inkjet Arrays

Agilent has adapted the inkjet printing technology of Hewlett Packard to the manufacturing of DNA microarrays on glass slides. There are two fundamen-

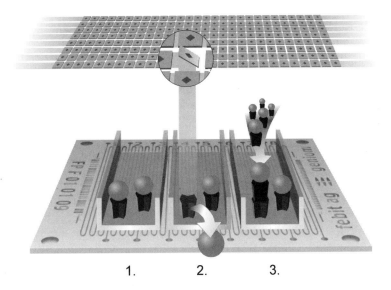

Fig. 1.8   Graphical illustration of the *in situ* synthesis of probes inside the Febit DNA processor. Shown are three enlargements of a microchannel, each illustrating one step in the synthesis. 1: the situation before synthesis. 2: selected positions are deprotected by controlling light illumination via a micromirror. 3: substrate is added to the microchannel and covalently attached to the deprotected positions. (See color plate. Copyright Febit AG. Used with permission.)

Table 1.3   Performance of the Febit Geniom One technology (Febit, 2002).

Starting material	10 μg total RNA
Detection limit	0.5 pM spiked transcript control
Number of probes per gene	4–10
Probe length	User selectable (10–60mers)
Feature size	34 μm by 34 μm
Probes per array	Minimum 6,000
Arrays per DNA processor	Up to eight
Sample throughput	About 80 samples per week

tally different approaches. Pre-synthesized oligos or cDNAs can be printed directly on a glass surface. These are called deposition arrays. Another approach uses solid-phase phosporamidite chemistry to build the oligos on the array surface one nucleotide at a time.

Table 1.4 Performance of bead arrays (Illumina, 2003).

Starting material	50-200 ng total RNA
Detection limit	0.15 pM
Number of probes per gene	4–10
Probe length	50 mers
Feature spacing	6 μm
Probes per array	1,500
Array matrix	96 samples

1.3.5 Bead Arrays

Illumina (www.illumina.com) has marketed a bead-based array system. Instead of controlling the location of each spot on a slide, they let small glass beads with covalently attached oligo probes self-assemble into etched substrates. A decoding step is then performed to read the location of each bead in the array before it is used to hybridize to fluorescently labeled sample.

1.3.6 Serial Analysis of Gene Expression (SAGE)

A technology that is both widespread and attractive because it can be run on a standard DNA sequencing apparatus is serial analysis of gene expression (SAGE) (Velculescu et al., 1995; Yamamoto et al., 2001). In SAGE, cDNA fragments called tags are concatenated by ligation and sequenced. The number of times a tag occurs, and is sequenced, is related to the abundance of its corresponding messenger. Thus, if enough concatenated tags are sequenced one can get a quantitative measure of the mRNA pool. Bioinformatics first enters the picture when one wishes to find the gene corresponding to a particular tag, which may be only 9 to 14 bp long. Each tag is searched against a database (Van Kampen et al., 2000; Margulies et al., 2000; Lash et al., 2000) to find one (or more) genes that match.

The steps involved in the SAGE methods can be summarized as follows (see also Figure 1.9):

- Extract RNA and convert to cDNA using biotinylated poly-T primer.

- Cleave with a frequently cutting (4 bp recognition site) restriction enzyme (anchoring enzyme).

- Isolate 3′-most restriction fragment with biotin-binding streptavidin-coated beads.

- Ligate to linker that contains a type IIS restriction site for and primer sequence.

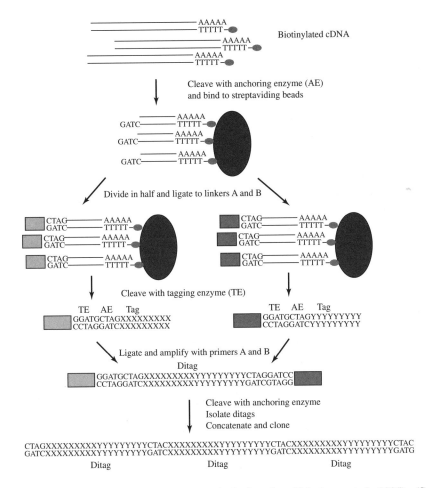

Fig. 1.9 Schematic overview of SAGE methods (based on Velculescu et al. 1995). (See color plate.)

- Cleave with tagging enzyme that cuts up to 20 bp away from recognition site.

- Ligate and amplify with primers complementary to linker.

- Cleave with anchoring enzyme, isolate ditags.

- Concatenate and clone.

- Sequence clones.

The analysis of SAGE data is similar to the analysis of array data described through out this book except that the statistical analysis of significance is different (Man et al., 2000; Lash et al., 2000; Audic et al., 1997).

1.4 PARALLEL SEQUENCING ON MICROBEAD ARRAYS

A conceptual merger of the SAGE technology and the microbead array technology is found in the massively parallel signature sequencing (MPSS) on microbead arrays, marketed by Lynx (www.lynxgen.com). cDNA is cloned into a vector together with a unique sequence tag that allows it to be attached to a microbead surface where an anti-tag is covalently attached. Instead of quantifying the amount of attached cDNA to each bead, the number of beads with the same cDNA attached is determined. This is done by sequencing 16–20 basepairs of the cDNA on each bead. This is done by a clever procedure of repeated ligation and restriction cycles intervened by fluorescent decoding steps. The result is the number of occurrences of each 16–20 basepair signature sequence, that can be used to find the identity of each cDNA in a database just as is done with SAGE.

Fig. 1.10 Megaclone bead arrays. cDNA attached to bead surface via tag-antitag hybridization. (From Lynx used with permission.)

1.4.1 Emerging Technologies

Nanomechanical cantilevers (McKendry et al., 2002) can detect the hybridization of DNA without any fluorescent labeling. The cantilevers are made of silicon, coated with gold, and oligonucleotide probes are attached. When target-probe hybridization occurs, the cantilever bends slightly and this can be detected by deflection of a laser beam. The amount of deflection is a function of the concentration of the target, so the measurement is quantitative. An alternative to laser beam detection is piezo-resistive readout from each cantilever. As the number of parallel cantilevers increases, this technology shows promise for sensitive, fast, and economic quantification of mRNA expression. The company Concentris (www.concentris.com) currently offers a commercial 8-cantilever array.

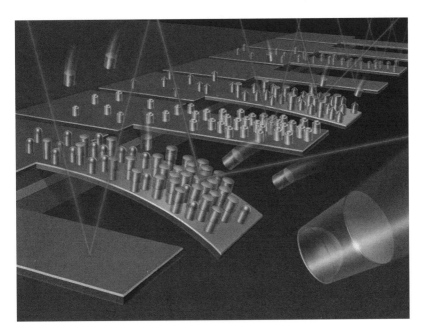

Fig. 1.11 Artist's illustration of array of 8 nanomechanical cantilevers. Binding of targets leads to bending that is detected by deflection of a laser beam. (Used with permission from Concentris.)

1.5 EXAMPLE: AFFYMETRIX VS. SPOTTED ARRAYS

Our lab has in a collaboration (Knudsen et al., 2001) performed both cDNA array analysis and Affymetrix chip analysis of human T-cells infected with Human Immunodeficiency Virus (HIV). Figure 1.12 shows mRNA extracted from the T-cells and visualized with a cDNA array. First, the mRNA was converted to cDNA and then it was labeled with a fluorochrome. We used a red fluorochrome for the mRNA that was extracted from the HIV-infected cells (experiment, sample, or treatment) and we used a green fluorochrome

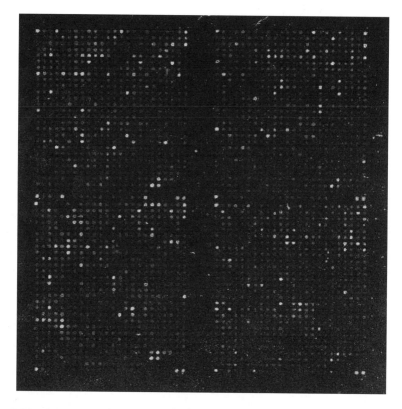

Fig. 1.12   cDNA microarray of genes affected by HIV infection. (See color plate.)

for mRNA that was extracted from the noninfected cells (control). Because we used different fluorochromes we could apply both sample and control to the same chip where we have already spotted probes for the genes we were interested in. Figure 1.12 shows such genes.

After hybridization and washing, the chip was scanned and the image processed by a computer. We can now deduce the ratio between the expression of each gene in HIV-infected cells and the gene in uninfected cells as the ratio between the intensity of red and green color. If the color is yellow, there is no change. If it is red, there is an upregulation; if it is green there is a downregulation (Figure 1.12). We can also estimate mRNA concentration from the intensity of the spot.

We took the same mRNA, processed it, and put it on an Affymetrix chip with 6800 human genes. Figure 1.13 shows part of the surface of this chip that probes for just one gene. Before putting it on the chip, the mRNA was

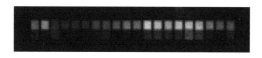

Fig. 1.13 Part of Affymetrix chip probing one gene affected by HIV infection.

converted to cDNA and an *in vitro* transcription step was used to amplify the amount of mRNA. After fragmentation and labeling with a fluorescent dye, the sample was hybridized to the chip surface where a total of 40 oligonucleotide probes of length 25 are used to detect the presence and concentration of each gene messenger. Twenty oligos are chosen from different areas of the gene. Each of these oligos is called a perfect match (PM). For each PM oligo there is an identical oligo with one mismatch at the center position of the 25-mer. This mismatch (MM) oligo is included to compensate for nonspecific hybridization as well as cross-hybridization.

In Figure 1.13 the PM oligos and MM oligos are shown on top of each other. Note that Affymetrix does not use a two-color system, so we have to run one chip for the sample and one chip for the control. The conditions for comparing those two chips will be described in a later section.

1.6 SUMMARY

Spotted arrays are made by deposition of a probe on a solid support. Affymetrix chips are made by light mask technology. The latter is easier to control and therefore the variation between chips is smaller in the latter technology. Spotted arrays offer more flexibility, however. Data analysis does not differ much between the two types of arrays. Digital micromirror technology combines the flexibility of the spotted arrays with the speed of the prefabricated Affymetrix chips.

Serial analysis of gene expression (SAGE) is yet another method for analyzing the abundance of mRNA by sequencing concatenated fragments of their corresponding cDNA. The number of times a cDNA fragment occurs in the concatenated sequence is proportional to the abundance of its corresponding messenger.

Figure 1.14 shows a schematic overview of how one starts with cells in two different conditions (with and without HIV virus, e.g.) and ends up with mRNA from each condition hybridized to a DNA array.

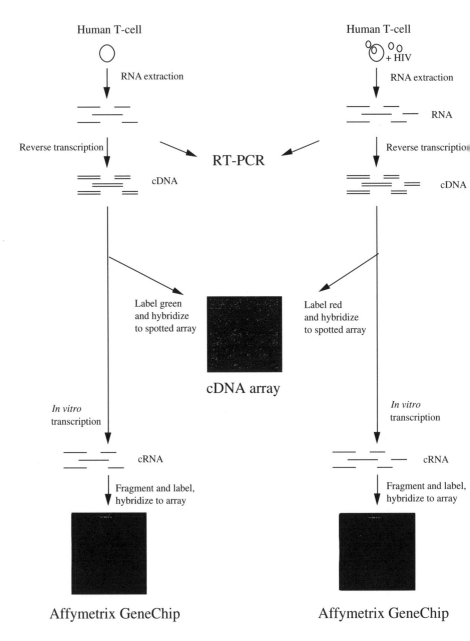

Fig. 1.14 Overview of methods for comparing mRNA populations in cells from two different conditions. (See color plate.)

Table 1.5 Overview of commercially available microarray technologies

	Factory synthesis	Customer synthesis
Affymetrix	mask-directed photolithography	
Agilent	ink-jet	
NimbleGen	micromirror photolithography	
Febit		micromirror photolithography
Spotted arrays	robot spotting	robot spotting
MPSS	cloning and sequencing	
Illumina Beads	oligos attached to beads	

1.7 FURTHER READING

Hood, L. (2002). Interview by The O'Reilly Network[1]

Knudsen, S., Nielsen, H. B., Nielsen, C., Thirstrup, K., Blom, N., Sicheritz-Ponten, T., Gautier, L., Workman, C., and Brunak, S. T-cell transcriptional responses to HIV infection in vitro. Unpublished.

Overview as well as details of Affymetrix technology

Affymetrix (1999). *GeneChip Analysis Suite*. User Guide, version 3.3.

Affymetrix (2000). *GeneChip Expression Analysis*. Technical Manual.

Baugh, L. R., Hill, A. A., Brown, E. L., and Hunter, C. P. (2001). Quantitative analysis of mRNA amplification by in vitro transcription. *Nucleic Acids Research* 29:E29.

Lipshutz, R. J., Fodor, S. P. A., Gingeras, T. R., and Lockhart, D. J. (1999). High density synthetic oligonucleotide arrays. *Nature Genetics Chipping Forecast* 21:20–24.

Lockhart, D. J., Dong, H., Byrne, M. C., Follettie, M. T., Gallo, M. V., Chee, M. S., Mittmann, M., Wang C., Kobayashi, M., Horton, H., and Brown, E. L. (1996). Expression monitoring by hybridization to high-density oligonucleotide arrays. *Nature Biotechnology* 14:1675–1680.

Wodicka, L., Dong, H., Mittmann, M., Ho, M. H., and Lockhart, D. J. (1997). Genome-wide expression monitoring in Saccharomyces cerevisiae. *Nature Biotechnology* 15:1359–1367.

---

[1]http://www.oreillynet.com/

Overview as well as details of spotted arrays

Bowtell, D., and Sambrook, J. (editors) (2002). DNA Microarrays: *A Molecular Cloning Manual*. New York: Cold Spring Harbor Laboratory Press.

Pat Brown lab web site.[2]

Schena, M. (2000). *Microarray Biochip Technology*. Sunnyvale, CA: Eaton.

Schena, M. (1999). *DNA microarrays: A practical approach* (Practical Approach Series, 205). Oxford: Oxford University Press.

Microarrays web site.[3] Includes protocols largely derived from the Cold Spring Harbor Laboratory Microarray Course manual.

Digital Micromirror Arrays

Albert, T. J., Norton, J., Ott, M., Richmond, T., Nuwaysir, K., Nuwaysir, E. F., Stengele, K. P., and Green, R. D. (2003). Light-directed 5′ → 3′ synthesis of complex oligonucleotide microarrays. *Nucleic Acids Research* 31(7):e35.

Baum, M., Bielau, S., Rittner, N., Schmid, K., Eggelbusch, K., Dahms, M., Schlauersbach, A., Tahedl, H., Beier, M., Guimil, R., Scheffler, M., Hermann, C., Funk, J. M., Wixmerten, A., Rebscher, H., Honig, M., Andreae, C., Buchner, D., Moschel, E., Glathe, A., Jager, E., Thom, M., Greil, A., Bestvater, F., Obermeier, F., Burgmaier, J., Thome, K., Weichert, S., Hein, S., Binnewies, T., Foitzik, V., Muller, M., Stahler, C. F., and Stahler, P. F. (2003). Validation of a novel, fully integrated and flexible microarray benchtop facility for gene expression profiling. *Nucleic Acids Research* 31(23):e151.

Beier, M., Baum. M,, Rebscher, H., Mauritz, R., Wixmerten, A., Stahler, C. F., Muller, M., and Stahler, P. F. (2002). Exploring nature's plasticity with a flexible probing tool, and finding new ways for its electronic distribution. *Biochem. Soc. Trans.* 30:78–82.

Beier, M., and Hoheisel, J. D. (2002). Analysis of DNA-microarrays produced by inverse in situ oligonucleotide synthesis. *J. Biotechnol.* 94:15–22.

Nuwaysir, E. F., Huang, W., Albert, T. J., Singh, J., Nuwaysir, K., Pitas, A., Richmond, T., Gorski, T., Berg, J. P., Ballin, J., McCormick, M.,

---

[2]http://brownlab.stanford.edu
[3]http://www.microarrays.org/

Norton, J., Pollock, T., Sumwalt, T., Butcher, L., Porter, D., Molla, M., Hall, C., Blattner, F., Sussman, M. R., Wallace, R. L., Cerrina, F., and Green, R. D. (2002). Gene expression analysis using oligonucleotide arrays produced by maskless photolithography. *Genome Research* 12:1749–1755.

Singh-Gasson, S., Green, R. D., Yue, Y., Nelson, C., Blattner, F., Sussman, M. R., and Cerrina F. (1999). Maskless fabrication of light-directed oligonucleotide microarrays using a digital micromirror array. *Nature Biotechnology* 17:974–978.

Inkjet arrays

Hughes, T. R., Mao, M., Jones, A. R., Burchard, J., Marton, M. J., Shannon, K. W., Lefkowitz, S. M., Ziman, M., Schelter, J. M., Meyer, M. R., Kobayashi, S., Davis, C., Dai, H., He, Y. D., Stephaniants, S. B., Cavet, G., Walker, W. L., West, A., Coffey, E., Shoemaker, D. D., Stoughton, R., Blanchard, A. P., Friend, S. H., and Linsley, P. S. (2001). Expression profiling using microarrays fabricated by an ink-jet oligonucleotide synthesizer. *Nature Biotechnology* 19:342–347.

van't Veer, L. J., Dai, H., van de Vijver, M. J., He, Y. D., Hart, A. A., Mao, M., Peterse, H. L., van der Kooy, K., Marton, M. J., Witteveen, A. T., Schreiber, G. J., Kerkhoven, R. M., Roberts, C., Linsley, P. S., Bernards, R., Friend, S. H. (2002). Gene expression profiling predicts clinical outcome of breast cancer. *Nature* 415:530–536.

Parallel Signature Sequencing

Brenner, S., Johnson, M., Bridgham, J., Golda, G., Lloyd, D. H., Johnson, D., Luo, S., McCurdy, S., Foy, M., Ewan, M., Roth, R., George, D., Eletr, S., Albrecht, G., Vermaas, E., Williams, S. R., Moon, K., Burcham, T., Pallas, M., DuBridge, R. B., Kirchner, J., Fearon, K., Mao, J., and Corcoran, K.. (2000). Gene expression analysis by massively parallel signature sequencing (MPSS) on microbead arrays. *Nature Biotechnology* 18:630–634.

Emerging Technologies

Marie, R., Jensenius, H., Thaysen, J. Christensen, C. B., and Boisen, A. (2002). Adsorption kinetics and mechanical properties of thiol-modied DNA-oligos on gold investigated by microcantilever sensors. *Ultramicroscopy* 91:29–36.

McKendry, R., Zhang, J., Arntz, Y., Strunz, T., Hegner, M., Lang, H. P., Baller, M. K., Certa, U., Meyer, E., Guntherodt, H. J., and Gerber, C. (2002). Multiple label-free biodetection and quantitative DNA-binding assays on a nanomechanical cantilever array. *Proc. Natl. Acad. Sci. USA* 99:9783–9788.

Serial analysis of gene expression

Audic, S., and Claverie, J. M. (1997). The significance of digital gene expression profiles. *Genome Res..* 7:986–995.

van Kampen, A. H., van Schaik, B. D., Pauws, E., Michiels, E. M., Ruijter, J. M., Caron, H. N., Versteeg, R., Heisterkamp, S. H., Leunissen, J. A., Baas, F., and van der Mee, M. (2000). USAGE: A web-based approach towards the analysis of SAGE data. *Bioinformatics.* 16:899–905.

Lash, A. E., Tolstoshev, C. M., Wagner, L., Schuler, G. D., Strausberg, R. L., Riggins, G. J., and Altschul, S. F. (2000). SAGEmap: A public gene expression resource. *Genome Res.* 10:1051–1060.

Man, M. Z., Wang, X., and Wang, Y. (2000). POWER_SAGE: Comparing statistical tests for SAGE experiments. *Bioinformatics.* 16:953–959.

Margulies, E. H., and Innis, J. W. (2000). eSAGE: Managing and analysing data generated with serial analysis of gene expression (SAGE). *Bioinformatics.* 16:650–651.

Velculescu, V. E., Zhang, L., Vogelstein, B., and Kinzler, K. W. (1995). Serial analysis of gene expression. *Science.* 270:484–487.

Yamamoto, M., Wakatsuki, T., Hada, A., and Ryo, A. (2001). Use of serial analysis of gene expression (SAGE) technology. *J Immunol. Methods.* 250:45–66. Review.

2

Overview of Data Analysis

Figure 2.1 presents an overview of the general data analysis methods presented in this book. In particular it shows the order in which to apply them and which methods to choose in different situations. For clarity, not all possible orders of analysis have been shown. For example, PCA and clustering can be performed after an ANOVA or t-test.

See Table 2.1 for where to find details of the individual methods.

Table 2.1 Section and page number for methods shown in Figure 2.1.

Method	Section	Page
Image analysis	3	25
Normalization	4.1	33
t-test	4.6	41
ANOVA	4.6	41
PCA	5.1	55
Clustering	6	63
Pathway analysis	8.1	85
Promoter analysis	7.2	78
Function prediction	7.1	77
Classification	10	101

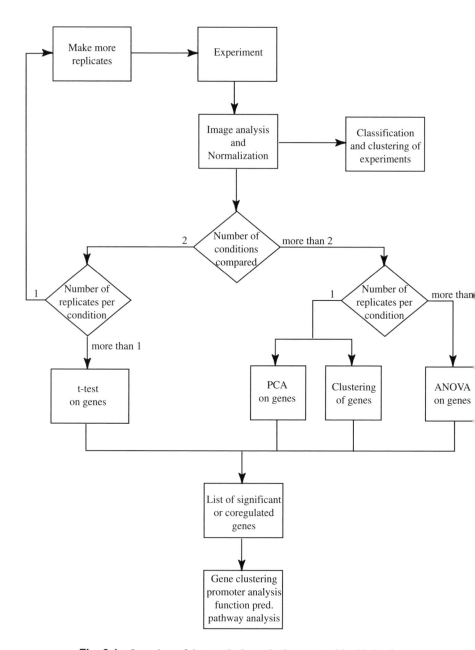

Fig. 2.1   Overview of data analysis methods presented in this book.

3

Image Analysis

Image analysis is an important aspect of microarray experiments. It can have a potentially large impact on subsequent analysis such as clustering or the identification of differentially expressed genes.

—Yang, 2001

Analysis of the image of the scanned array seeks to extract an intensity for each spot or feature on the array. In the simplest case, we seek one expression number for each gene. The analysis can be divided into several steps (Yang et al., 2001):

- Gridding

- Segmentation

- Intensity extraction

- Background correction

This chapter will first describe the basic concepts of image analysis and then list a number of software packages available for the purpose.

3.1 GRIDDING

Whether you have a scanned image of a spotted array or an image of an Affymetrix GeneChip, you need to identify each spot or feature. That is accomplished by aligning a grid to the spots, because the spots are arranged in a grid of columns and rows. For photolithographically produced chips this may be easier than for robot spotted arrays, where more variation in the grid is possible. For the latter a manual intervention may be necessary to make sure that all the spots have been correctly identified.

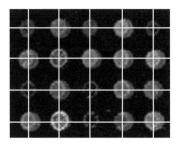

Fig. 3.1   Aligning a grid to identify the location of each spot.

3.2 SEGMENTATION

Once the spots have been identified, they need to be separated from the background. The shape of each spot has to be identified. The simplest assumption is that all spots are circular of constant diameter. Everything inside the circle is assumed to be signal and everything outside is assumed to be background. This simple assumption rarely holds, and therefore most image analysis software includes some more advanced segmentation method. Adaptive circle segmentation estimates the diameter separately for each spot. Adaptive shape segmentation does not assume circular shape of each spot and instead tries to find the best shape to describe the spot. Finally, the histogram method analyzes the distribution of pixel intensities in and around each spot to determine which pixels belong to the spot and which pixels belong to the background.

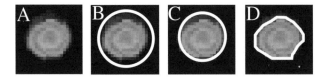

Fig. 3.2 Illustration of segmentation methods. A: Image before segmentation, B: Fixed circle segmentation, C: Adaptive circle segmentation, D: Adaptive shape segmentation.

3.3 INTENSITY EXTRACTION

Once the spot has been separated from the surrounding background, an intensity has to be extracted for each spot and potentially for each surrounding background. Typical measures are the mean or median intensity of all pixels within the spot.

3.4 BACKGROUND CORRECTION

On some array images, a slight signal is seen in the area that is in between spots. This is a background signal and it can be subtracted from the spot intensity to get a more accurate estimate of the biological signal from the spot. There are some problems associated with such a background correction, however. First, it does not necessarily follow that the background signal is added to the spot signal. In other words, some spots can be seen with lower intensity than the surrounding background. In that case, the surrounding background should clearly not be subtracted from the spot.

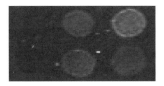

Fig. 3.3 Illustration of "ghost" where the background has higher intensity than the spot. In this case, subtracting background from spot intensity may be a mistake.

Second, a local estimation of background is necessarily associated with some noise. Subtracting such a noisy signal from a weak spot signal with noise will result in a number with even more noise. For weakly expressed genes this noise increase can negatively affect the following statistical analysis for differential expression.

The effect of not subtracting a background is that the absolute values may be slightly higher and that fold changes may be underestimated slightly. On balance, we choose not to subtract any local background but we do subtract a globally estimated background. This can, for example, be the second or third percentile of all the spot values. This is similar to the approach used by Affymetrix GeneChip software, where the image is segmented into 16 squares, and the average of the lower 2% of feature intensities for each block is used to calculate background. This background intensity is subtracted from all features within a block.

That leaves the issue of spatial bias on an array. This topic is usually considered under normalization. We have investigated spatial bias both for spotted arrays and Affymetrix arrays and found it to be significant in spotted arrays (Workman et al., 2002). We define spatial bias as an overall trend of fold changes that vary with the location on the surface (see Figure 3.4). Such a bias can be removed with Gaussian smoothing (Workman, 2002). In essence, the local bias in fold change is calculated in a window and subtracted from the observed fold change.

3.5 SOFTWARE

3.5.1 Free Software for Array Image Analysis

(From http://ihome.cuhk.edu.hk/~b400559/arraysoft_image.htm)

- **Dapple**
  Washington University.
  http://www.cs.wustl.edu/%7Ejbuhler//research/dapple/

- **F-Scan**
  National Institutes of Health.
  http://abs.cit.nih.gov/fscan/

- **GridGrinder**
  Corning Inc.
  http://gridgrinder.sourceforge.net/

- **Matarray**
  Medical College of Wisconsin.
  http://www.mcw.edu/display/router.asp?docid=530

- **P-Scan**
  National Institutes of Health.
  http://abs.cit.nih.gov/pscan/index.html

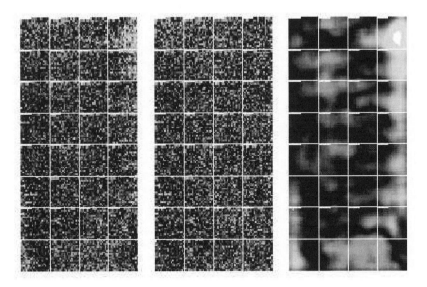

Fig. 3.4 Spatial effects on a spotted array. The blue-yellow color scale (see color plate) indicates fold change between the two channels. A spatial bias is visible (left). Gaussian smoothing captures the bias (right), which can the be removed by subtraction from the image (center). From Workman (2002).

- **ScanAlyze**
  Lawrence Berkeley National Lab.
  http://rana.lbl.gov/EisenSoftware.htm

- **Spotfinder**
  The Institute for Genomic Research.
  http://www.tigr.org/software/tm4/spotfinder.html

- **UCSF Spot**
  University of California, San Francisco.
  http://jainlab.ucsf.edu/Downloads.html

3.5.2 Commercial Software for Array Image Analysis

(From http://ihome.cuhk.edu.hk/~b400559/arraysoft_image.htm)

- **AIDA Array Metrix**
  Raytest GmbH.
  http://www.raytest.de

- **ArrayFox**
  Imaxia Corp.
  http://www.imaxia.com/products.htm

- **ArrayPro**
  Media Cybernetics, Inc.
  http://www.mediacy.com/arraypro.htm

- **ArrayVision**
  Imaging Research Inc.
  http://www.imagingresearch.com/products/ARV.asp

- **GenePix Pro**
  Axon Instruments, Inc.
  http://www.axon.com/GN_GenePixSoftware.html

- **ImaGene**
  BioDiscovery, Inc.
  http://www.biodiscovery.com/imagene.asp

- **IconoClust**
  CLONDIAG Chip Technologies GmbH.
  http://www.clondiag.com/products/sw/iconoclust/

- **Microarray Suite**
  Scanalytics, Inc.
  http://www.scanalytics.com/product/microarray/index.shtml

- **Koadarray**
  Koada Technology.
  http://www.koada.com/koadarray/

- **Lucidea**
  Amersham Biosciences.
  http://www1.amershambiosciences.com/

- **MicroVigene**
  VigeneTech, Inc.
  http://www.vigenetech.com/product.htm

- **Phoretics Array**
  Nonlinear Dynamics.
  http://www.phoretix.com/products/array_products.htm

- **Quantarray**
  PerkinElmer, Inc.
  http://las.perkinelmer.com/

- **Spot**
  CSIRO Mathematical and Information Sciences.
  http://experimental.act.cmis.csiro.au/Spot/index.php

3.6 SUMMARY

The software that comes with your scanner is usually a good start for image analysis. If possible, use a global background correction instead of subtracting a locally estimated background from each spot. It is always a good idea to look at the image of a chip to observe any visible defects, bubbles, or clear spatial bias.

3.7 FURTHER READING

Yang, Y. H., Buckley, M. J., and Speed, T. P. (2001). Analysis of cDNA microarray images. *Briefings in Bioinformatics* 2(4):341–349.

Yang, Y. H., Buckley, M. J., Dudoit, S., and Speed, T. P. (2001). Comparison of methods for image analysis on cDNA microarray data. Technical report #584, Department of Statistics, University of California, Berkeley.

Workman, C., Jensen, L. J., Jarmer, H., Berka, R., Saxild, H. H., Gautier, L., Nielsen, C., Nielsen, H. B., Brunak, S, and Knudsen, S. (2002). A new non-linear normalization method for reducing variance between DNA microarray experiments. *Genome Biology* 3(9):0048.[1]

---

[1] Software available in affy package of Bioconductor http://www.bioconductor.org

4

Basic Data Analysis

In gene expression analysis, technological problems and biological variation make it difficult to distinguish signal from noise. Once we obtain reliable data, we look for patterns and need to determine their significance.

—Vingron, 2001

4.1 NORMALIZATION

Microarrays are usually applied to the comparison of gene expression profiles under different conditions. That is because most of the biases and limitations that affect absolute measurements do not affect relative comparisons. There are a few exceptions to that. One is that you have to make sure that what you are comparing is really comparable. The chips have to be the same under the different conditions, but also the amount of sample applied to each chip has to be comparable. An example will illustrate this. Figure 4.1 shows a comparison between two chips where the same labeled RNA has been added to both. Ideally, all the intensity measurements on one chip should match those on the other, all points should lie on the diagonal. They do not, and they reveal both random and systematic bias. The systematic bias is revealed by a deviation from the diagonal that increases with intensity. This is a systematic bias that is signal dependent. It becomes even more pronounced when we plot the logarithm of the ratio versus the logarithm of the intensity (Figure 4.1B). This is often referred to as an M vs. A plot or MVA plot and it is often used to identify signal dependent biases.

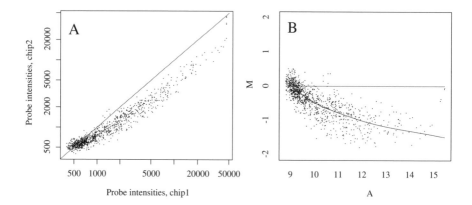

Fig. 4.1 Comparison of the probe intensities between two Affymetrix chips with the same labeled RNA applied. A: Log of intensity versus log of intensity. B: Log of ratio ($M =$ log(chip1/chip2)) versus average log intensity ($A = $ (log chip1 + log chip2)/2). The curve in B shows a lowess (locally weighted least squares fit) applied to the data.

We cannot remove the random bias (we will deal with it later by using replicates), but we can remove systematic bias. This is often referred to as normalization. Normalization is based on some assumptions that identify reference points.

4.1.1 One or More Genes Assumed Expressed at Constant Rate

These genes are referred to as housekeeping control genes. Examples include the GAPDH gene. Multiply all intensities by a constant until the expression of the control gene is equal in the arrays that are being compared. For arrays with few genes this is often the only normalization method available. This is, however, a linear normalization that does not remove the observed signal dependent nonlinearity. In the MVA plot of Figure 4.1B it amounts to addition of a constant to the ratio to yield the normalized data of Figure 4.2A. The systematic bias is still present. It would lead you to conclude that weakly expressed genes are upregulated on chip 1 relative to chip 2, whereas highly expressed genes are downregulated on chip 1 relative to chip 2. This conclusion we know is false, because be put the same RNA on both chips. If you have more control genes with different intensity you can draw a normalization curve by fitting a curve through them. That makes the normalization signal-dependent similar to what is seen in Figure 4.2B.

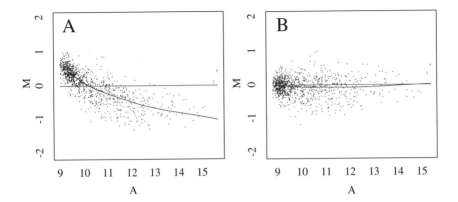

Fig. 4.2 Comparison of linear normalization (A) with signal dependent normalization (B).

4.1.2 Sum of Genes is Assumed Constant

This assumes that the total messenger RNA in a cell is constant which may hold true if you are using a large transcriptome chip with thousands of genes. It is the approach used by Affymetrix in the GeneChip software. All intensities on one chip are multiplied by a constant until they have the same sum or average as the chip you want to compare to. The problem is that this is still a linear normalization, that does not remove the signal dependent bias (Figure 4.2A). If your array contains only a few hundred genes that were selected for their participation in a certain process, the assumption of constant sum may not be a good one. Then it may be better to rely on one or more housekeeping control genes.

4.1.3 Subset of Genes is Assumed Constant

Li and Wong (2001b) have proposed a method whereby the constant control genes are not known *a priori*, but are instead identified as genes whose intensity rank does not differ more than a threshold value between two arrays. This invariant set is defined iteratively and used to draw a normalization curve that is signal dependent. In practice, this method works extremely well, and it has compared well to all other normalization methods developed.

4.1.4 Majority of Genes Assumed Constant

We have developed a signal dependent normalization method, qspline (Workman, 2002), that assumes that the overall intensity distributions between two

arrays should be comparable. That means that the quantiles[1] of the distributions, plotted in a quantile-quantile plot, should lie on the diagonal. If they
do not, they form a normalization curve that is signal dependent. After normalization, the scatter point smoothing line is close to a straight line through
zero in the MVA plot (Figure 4.2B). We have compared our method, qspline,
to the methods of Li-Wong (2001b) and Irizarry (2003) and found them to
behave very similarly on the datasets we have tested. For spotted arrays, we
have found our qspline method to perform slightly better on one dataset than
lowess as proposed by Yang (2002).

Even if the assumption does not hold, and less than a majority of genes is
constant, the normalization still works provided that the number of upregulated genes roughly equals the number of downregulated genes and provided
that there is no signal-dependent bias in up- or downregulation.

4.1.5 Spike Controls

If none of the above assumptions seem applicable to your experiment, there
is one choice of last resort. You can add a spiked control to your mRNA
preparation. The idea is to measure the amount of mRNA or total RNA
extracted from the cell, and then add a known transcript of known concentration to the pool. This spiked transcript is then assumed to be amplified
and labeled the same way as the other transcripts and detected with a unique
probe on the array. The spiked transcript must not match any gene in your
RNA preparation, so for a human RNA preparation an *E. coli* gene could be
used. After scanning the array you multiply all measurements on one array
until the spiked control matches that on the other array.

This approach has the limitation that it results in a linear normalization that
does not correct signal-dependent bias (unless you use many spiked control
genes with different concentration). Finally, it is only as accurate as the
accuracy of measuring the total amount of RNA and the accuracy of adding
an exact amount of spiked transcript.

4.2 DYE BIAS, SPATIAL BIAS, PRINT TIP BIAS

The above mentioned global normalization methods (in particular the signal-
dependent ones) work well for factory-produced oligonucleotide arrays such
as Affymetrix GeneChips. For arrays spotted with a robot, however, there
may be substantial residual bias after a global normalization. First, there is a
dye bias. The dyes Cy3 and Cy5 have different properties that are revealed if
you label the same sample with both dyes and then plot the resulting intensities

---

[1]A quantile has a fixed number of genes below it in intensity. The first quantile could have 1% of the
genes below it in intensity, the second quantile have 2% of the genes below it in intensity, and so on.

against each other. Such a plot reveals that the bias is signal dependent. For that reason, the signal-dependent normalization methods mentioned above will also remove the dye bias, and after signal-dependent normalization you can directly compare Cy3 channels to Cy5 channels. If you are unable to use a signal dependent normalization method, a typical approach to removing dye bias is to use a dye swap, which means that you label each sample both with Cy3 and Cy5 and then take the ratio of the averages of each sample. Note, however, that dye swap normalization does not remove signal dependent bias beyond the dye bias.

Spatial bias was dealt with in the Chapter 3. Removing it requires knowledge of the layout of the array.

Finally it is often possible to observe a print tip bias. The spots on the array are usually not printed with the same printing tip. Instead, several tips are used to print in parallel. So the spots on the array are divided into groups where each group of spots have been made with the same tip. When you compare the average log ratio within the print tip groups, you can observe differences between print tip groups. These can be due to a non-random order in which genes are printed, or they can be due to spatial biases or they can be due to true physical differences between the print tips. To remove the print tip bias is relatively straightforward. First you perform a linear normalization of each print tip group until their average signal ratio is equal. Then you perform a global signal dependent normalization to remove any signal dependent bias.

4.3 EXPRESSION INDICES

For spotted arrays using only one probe for each gene you can calculate fold changes after normalization. For Affymetrix GeneChips and other technologies relying on several different probes for each gene it is necessary to condense these probes into a single intensity for each gene. This we refer to as an expression index.

4.3.1 Average Difference

Affymetrix, in the early version (MAS 4.0) of their software, calculated an Average Difference between probe pairs. A probe pair consists of a perfect match (PM) oligo and a mismatch (MM) oligo for comparison. The mismatch oligo differs from the perfect match oligo in only one position and is used to detect nonspecific hybridization. Average Difference was calculated as follows:

$$\text{AvgDiff} = \frac{\sum_N (\text{PM} - \text{MM})}{N},$$

where N is the number of probe pairs used for the calculation (probe pairs which deviate by more than 3 standard deviations from the mean are *excluded* from the calculation). If the AvgDiff number is negative or very small, it means that either the target is absent or there is nonspecific hybridization. Affymetrix calculates an Absolute Call based on probe statistics: Absent, Marginal, or Present (refer to the Affymetrix manual for the decision matrix used for making the Absolute Call).

4.3.2 Signal

In a later version of their software (MAS 5.0), Affymetrix has replaced AvgDiff with a Signal, which is calculated as

$$\text{Signal} = \text{Tukeybiweight}[\log(PM_n - CT_n)],$$

where Tukey biweight is a robust estimator of central tendency. To avoid negative numbers when subtracting the mismatch, a number CT is subtracted that can never be larger than PM. Note, however, that this could affect the normality assumption often used in downstream statistical analysis (Giles and Kipling, 2003).

4.3.3 Model-Based Expression Index

Li and Wong (2001a, b) instead calculate a weighted average difference:

$$\tilde{\theta} = \frac{\sum_N (\text{PM}_n - \text{MM}_n)\phi_n}{N},$$

where ϕ_n is a scaling factor that is specific to probe pair $\text{PM}_n - \text{MM}_n$ and is obtained by fitting a statistical model to a series of experiments. This model takes into account that probe pairs respond differently to changes in expression of a gene and that the variation between replicates is also probe-pair dependent. Li and Wong have also shown that the model works without the mismatches (MM) and then usually has lower noise than when mismatches are included. Software for fitting the model (weighted average difference and weighted perfect match), as well as for detecting outliers and obtaining estimates on reliability is available for download.[2] Lemon and coworkers have compared the Li-Wong model to Affymetrix' Average Difference and found it to be superior in a realistic experimental setting (Lemon et al., 2001). Note that model parameter estimation works best with 10 to 20 chips.

4.3.4 Robust Multiarray Average

Irizarry et al. (2003) have published a Robust Multiarray Average that also reduces noise by omitting the information present in the mismatch probes of Affymetrix GeneChips:

---

[2]http://www.dchip.org

$$RMA = \text{Medianpolish}[\log PM_n - \alpha_n)],$$

where Median polish is a robust estimator of central tendency and α is a scaling factor that is specific to probe PM_n and is obtained by fitting a statistical model to a series of experiments.

4.3.5   Position Dependent Nearest Neighbor Model

All the above expression indices are a statistical treatment of the probe data that assume that the performance of a probe can be estimated from the data. While this is true, there may be an even more reliable way of estimating probe performance: based on thermodynamics. The field of probe design for microarrays has been hampered by an absence of good thermodynamic models that accurately describe hybridization to an oligo attached to an array surface. Zhang et al. (2003a, b) have developed just that. They extend the nearest neighbor energy model, that works well for oligonucleotides in solution, with a position term that takes into account whether a nucleotide (or pair of nucleotides) is at the center of a probe or near the array surface or near the free end of the probe. They can use real array data to estimate the parameters of this model, and the resulting model works quite well at modeling the sequence dependent performance of each probe. As such it can be used for condensing the individual probe measures into one number for each gene.

4.4   DETECTION OF OUTLIERS

Outliers in chip experiments can occur at several levels. You can have an entire chip that is bad and consistently deviates from other chips made from the same condition or sample. Or you can have an individual gene on a chip that deviates from the same gene on other chips from the same sample. That can be caused by image artifacts such as hairs, air bubbles, precipitation, and so on. Finally, it is possible that a single probe, due to precipitation or other artifact, is perturbed.

How can you detect outliers in order to remove them? Basically, you need a statistical model of your data. The simplest model is equality among replicates. If one replicate (chip, gene, or probe) deviates several standard deviations from the mean, you can consider it an outlier and remove it. The t-test measures standard deviation and gives genes where outliers are present among replicates a low significance (See Section 4.6).

More advanced statistical models have been developed that also allow for outlier detection and removal (Li and Wong, 2001a, b).

4.5 FOLD CHANGE

Having performed normalization (and, if necessary, expression index condensation) you should now be able to compare the expression level of any gene in the sample to the expression level of the same gene in the control. The next thing you want to know is: How many fold up- or downregulated is the gene, or is it unchanged?

The simplest approach to calculate fold change is to divide the expression level of a gene in the sample by the expression level of the same gene in the control. Then you get the fold change, which is 1 for an unchanged expression, less than 1 for a down-regulated gene, and larger than 1 for an up-regulated gene. The definition of fold change will not make any sense if the expression value in the sample or in the control is zero or negative. Early Average Difference values from Affymetrix sometimes were, and a quick-and-dirty way out of this problem was to set all Average Difference values below 20 to 20. This was the approach usually applied.

The problem with fold change emerges when one takes a look at a scale. Up-regulated genes occupy the scale from 1 to infinity (or at least 1000 for a 1000-fold up-regulated gene) whereas all down-regulated genes only occupy the scale from 0 (0.001 for a 1000-fold down-regulated gene) to 1. This scale is highly asymmetric.

The Affymetrix GeneChip software (early version MAS 4.0) calculates fold change in a slightly different way, which does stretch out that scale to be symmetric:

$$\text{AffyFold} = \frac{\text{Sample} - \text{Control}}{\min(\text{Sample, Control})} + \begin{cases} +1 & \text{if Sample} > \text{Control} \\ -1 & \text{if Sample} < \text{Control} \end{cases},$$

where Sample and Control are the AvgDiffs of the sample and the control, respectively. For calculation of fold change close to the background level, consult the Affymetrix manual.

This function is discontinuous and has no values in the interval from -1 to 1. Up-regulated genes have a fold change greater than 1 and down-regulated genes have a fold change less than -1. But the scale for down-regulated genes is comparable to the scale for up-regulated genes

Both the fold change and Affyfold expressions are intuitively rather easy to grasp and deal with, but for further computational data analysis they are not useful, either because they are asymmetric or because they are discontinuous. For further data analysis you need to calculate the logarithm of fold change. Logfold, as we will abbreviate it, is undefined for the Affymetrix fold change, but can be applied to the simple fold change provided that you have taken the precaution to avoid values with zero or negative expression.

It is not important whether you use the natural logarithm ($\log_e$), base 2 logarithm ($\log_2$), or base 10 logarithm ($\log_{10}$).

4.6 SIGNIFICANCE

If you have found a gene that is twofold up-regulated ($\log_{10}$ fold is 0.3), then how do you know whether this is not just a result of experimental error? You need to determine whether or not a twofold regulation is *significant*. There are many ways to estimate significance in chip experiments. Basically, to assess experimental error you have to repeat the experiment and measure the variation. If you repeat both control and sample, you can use a *t*-test to determine whether the expression of a particular gene is significantly different between control and sample.

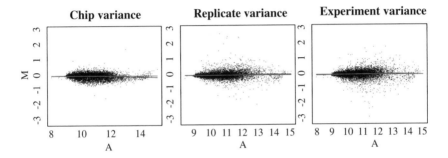

Fig. 4.3   Observed fold changes when comparing (left) chips with the same labeled mRNA, (middle) chips with mRNA preparations from two replicate cultures, (right) chips from two different experimental conditions. Plots show log of ratio ($M = \log(\text{chip1}/\text{chip2})$) versus average log intensity ($A = (\log \text{chip1} + \log \text{chip2})/2$) for all genes.

The *t*-test looks at the mean and variance of the sample and control distributions and calculates the probability that the observed difference in mean occurs when the null hypothesis is true[3]. When using the *t*-test it is often assumed that there is equal variance between sample and control. That allows the sample and control to be pooled for variance estimation. If the variance cannot be assumed equal you can use Welch's *t*-test which assumes unequal variances of the two populations.

When using the *t*-test for analysis of microarray data, it is often a problem that the number of replicates is low. The lower the number of replicates, the more difficult it will be to estimate the variance. For only two replicates it

---

[3]The null hypothesis states that the mean of the two distributions is equal. Hypothesis testing allows us to calculate the probability of finding the observed data when the hypothesis is true. To calculate the probability we make use of the normal distribution. When the probability is low, we reject the null hypothesis.

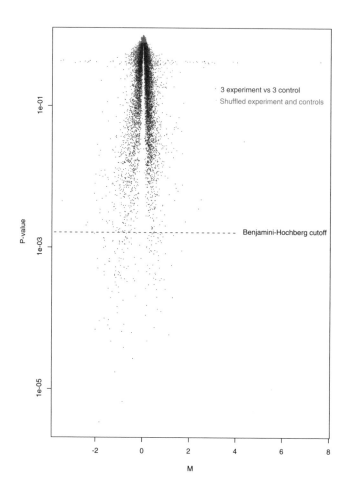

Fig. 4.4 Volcano plot showing relationship between *P*-value and logfold change (M) (Wolfinger, 2001). M is the average of three replicates of each condition. For comparison a random permutation of the data has been included (green points, see color plate).

becomes almost impossible. There are several solutions to this. The simplest solution is to take fold change into account for experiments with low number of replicates (say less than 3), and not consider genes that have less than 2 fold change in expression. This will guard against low P-values that arise from underestimation of variance, and is similar to the approach used in SAM where a constant is added to the gene-specific variance (Tusher et al., 2001). The other possibility is to base variance estimation not only on a single gene measurement, but to include variance estimates from the whole population. Several methods have been developed for this purpose (Kerr et al., 2001; Baldi and Long, 2001; Lonnstedt and Speed, 2001).

The relationship between P-value and fold change is illustrated in a so-called volcano plot (Figure 4.4). You can observe large fold changes that are not significant (they are most likely due to outliers). Among the significant genes (low P-value), you observe genes with both large and small logfold change. For comparison, a permutation of the data is included as well. Here, samples and controls have been shuffled before repeating the analysis. This shows the amount of noise in the data, and what fold changes and P-values can be observed in a random permutation of the data. When you choose a cutoff in P-value, you can see how many data points from the random permutation exceed that cutoff. This is an estimate of your false positive rate.

4.6.1 Multiple Conditions

If you have more than two conditions, the t-test may not be the method of choice, because the number of comparisons grow if you perform all possible comparisons between conditions. The method analysis of variance (ANOVA) will, using the F distribution, calculate the probability of finding the observed differences in means between more than two conditions when the null hypothesis is true (when there is no difference in means).

Software for running the t-test and ANOVA will be discussed in Section 14.5. A web-based method for t-test is available[4] (Baldi and Long, 2001). Baldi and Long (2001) recommend using the t-test on log-transformed data.

4.6.2 Nonparametric Tests

Both the t-test and ANOVA assume that your data follow the normal distribution. Although both methods are robust to moderate deviations from the normal distribution[5], alternative methods exist for assessing significance without assuming normality. The Wilcoxon/Mann-Whitney rank sum test

---

[4]http://visitor.ics.uci.edu/genex/cybert/
[5]Giles and Kipling (2003) have demonstrated that deviations from the normal distribution are small for most microarray data, except when using Affymetrix' MAS 5.0 software

will do the same without using the actual expression values from the experiment, only their rank relative to each other.

When you rank all expression levels from the two conditions, the best separation you can have is that all values from one condition rank higher than all values from the other condition. This corresponds to two non overlapping distributions in parametric tests. But since the Wilcoxon test does not measure variance, the significance of this result is limited by the number of replicates in the two conditions. It is for this reason that you may find that the Wilcoxon test for low numbers of replication gives you a poor significance and that the distribution of P-values is rather granular.

4.6.3 Correction for Multiple Testing

For all statistical tests that calculate a P-value, it is important to consider the effect of multiple testing as we are looking at not just one gene but thousands of genes. If a P-value of 0.05 tells you that you have a probability of 5% of making a type I error (false positive) on one gene, then you expect 500 type I errors (false positive genes) if you look at 10.000 genes. That is usually not acceptable. You can use the Bonferroni correction to reduce the significance cutoff to a level where you again have only 5% probability of making one or more type I errors among all 10.000 genes. This new cutoff is $0.05/10.000 = 5 \cdot 10^{-6}$. That is a pretty strict cutoff, and for many purposes you can live with more type I errors than that. Say you end up with a list of 100 significant genes and you are willing to accept 5 type I errors (false positives) on this list. Then you are willing to accept a *False Discovery Rate (FDR)* of 5%. Benjamini and Hochberg (1995) have come up with a method for controlling the FDR at a specified level. After ranking the genes according to significance (P-value) and starting at the top of the list, you accept all genes where

$$P \leq \frac{i}{m}q,$$

where i is the number of genes accepted so far, m is the total number of genes tested and q is the desired FDR. For $i > 1$ this correction is less strict than a Bonferroni correction.

The False Discovery Rate can also be assessed by permutation. If you permute the measurements from sample and control and repeat the t-test for all genes, you get an estimate of the number of type I errors (false positives) that can be expected at the chosen cutoff in significance. When you divide this number by the number of genes that pass the t-test on the unpermuted data, you get the FDR. This is the approach used in the software SAM (Tusher et al., 2001).

If no genes in your experiment pass the Bonferroni or Benjamini-Hochberg corrections, then you can look at those that have the smallest P-value. When

Table 4.1   Expression readings of four genes in six patients.

| Gene | Patient | | | | | |
	N_1	N_2	A_1	A_2	B_1	B_2
a	90	110	190	210	290	310
b	190	210	390	410	590	610
c	90	110	110	90	120	80
d	200	100	400	90	600	200

you multiply their P-value by the number of genes in your experiment, you get an estimate of the number of false positives. Take this false positive rate into account when planning further experiments.

4.6.4   Example I: t-Test and ANOVA

A small example using only four genes will illustrate the t-test and ANOVA. The four genes are each measured in six patients, which fall into three categories: normal (N), disease stage A, and disease stage B. That means that each category has been *replicated* once (Table 4.1).

We can perform a t-test (see Section 14.5 for details) to see if genes differ significantly between patient category A and patient category B (Table 4.2). But you should be careful performing a t-test on as little as two replicates in real life. This is just for illustration purposes.

Gene b is significantly different at a 0.05 level, even after multiplying the P-value by four to correct for multiple testing. Gene a is not significant at a 0.05 level after Bonferroni correction, and genes c and d have a high probability of being unchanged. For gene d that is because, even though an increasing trend is observed, the variation within each category is too high to allow any conclusions.

Table 4.2 t-test on difference between patient categories A and B.

| Gene | Patient | | | | P-value |
	A_1	A_2	B_1	B_2	
a	190	210	290	310	0.019
b	390	410	590	610	0.005
c	110	90	120	80	1.000
d	400	90	600	200	0.606

Table 4.3 ANOVA on difference between patient categories N, A and B.

Gene	N_1	N_2	A_1	A_2	B_1	B_2	P-value
			Patient				
a	90	110	190	210	290	310	0.0018
b	190	210	390	410	590	610	0.0002
c	90	110	110	90	120	80	1.0000
d	200	100	400	90	600	200	0.5560

Table 4.4 Effect of number of replicates on Type I (FP) and II (FN) errors in t-test.

	2	3	4	5	6
True positives	23	144	405	735	1058
False positives	8	18	29	45	0
False negatives	1035	914	653	323	0

Number of replicates of each condition

If we perform an ANOVA instead, testing for genes that are significantly different in at least one of three categories, the picture changes slightly (Table 4.3).

In the ANOVA, both gene a and b are significant at a 0.01 level even after Bonferroni correction. So taking all three categories into account increases the power of the test relative to the t-test on just two categories.

4.6.5 Example II: Number of Replicates

If replication is required to determine the significance of results, how many replicates are required? An example will illustrate the effect of the number of replicates. We have performed six replicates of each of two conditions in a *Saccharomyces cerevisiae* GeneChip experiment (Piper et al., 2002). Some of the replicates have even been performed in different labs. Assuming the results of a t-test on this data set to be the correct answer, we can ask: How close would five replicates have come to that answer? How close would four replicates have come to that answer? We have performed this test (Piper et al., 2002) and Table 4.4 shows the results. For each choice of replicates, we show how many false positives (Type I errors) we have relative to the correct answer and how many false negatives (Type II errors) we have relative to the correct answer. The number of false positives in the table is close to the number we have chosen with our cutoff in the Bonferroni corrected t-test (a 0.005 cutoff at 6383 genes yields 32 expected false positives). The number of false negatives, however, is greatly affected by the number of replicates.

Table 4.5 Effect of number of replicates on Type I and II errors in SAM (Tusher et al., 2001).

	Number of replicates of each condition				
	2	3	4	5	6
True positives	27	165	428	748	1098
False positives	3	4	14	27	0
False negatives	1071	933	670	350	0

Table 4.5 shows that the t-test performs almost as well as SAM[6] (Tusher et al., 2001), which has been developed specifically for estimating the false positive rate in DNA microarray experiments based on permutations of the data.

While this experiment may not be representative, it does illustrate two important points about the t-test. You can control the number of false positives even with very low numbers of replication. But you lose control over the false negatives as the number of replications go down.

So how many replicates do you have to perform to avoid any false negatives? That depends mainly on two parameters. How large is the variance between replicates and how small a fold change do you wish to detect. Given those it is possible to calculate the number of replicates needed to achieve a certain power (1 minus the false negative rate) in the t-test[7].

4.7 MIXED CELL POPULATIONS

The analysis presented above assumes that we are looking at pure cell populations. In cell cultures we are assuming that all cells are identical, and in tissue samples we are assuming that the tissue is homogeneous. The degree to which these assumptions are true vary from experiment to experiment. If you have isolated a specific cell type from blood, there may still be a mixture of subtypes within this population. A tissue sample may contain a combination of tumor and normal cells. A growing cell culture contains cells in different stages of the cell cycle.

If the proportion of subtypes is constant throughout the experiment, a standard analysis can be applied. You just have to remember that any signal arising from a single subtype will be diluted by the presence of other subtypes.

If the proportion of subtypes vary in the experiment, it may be possible to resolve mathematically the proportions and estimate the expression of individual subtypes within the population. But the mathematical procedures

---

[6]Software available for download at http://www-stat.stanford.edu/~tibs/SAM/index.html
[7]For example by using the power.t.test function of the R package available from www.r-project.org

for doing this again rest on a number of assumptions. If you assume that each cell subtype has a uniform and unchanging expression profile, and that the only thing that changes in your experiment is the ratio between cell subtypes, the problem becomes a simple mathematical problem of solving linear equations (Lu et al., 2003). First you need to obtain the expression profile of each of the pure cell subtype in an isolated experiment. Then you find the linear combination of the pure profile that best fits the data of the mixed cell population experiments. This gives the proportions of the individual cell subtypes.

Another way of separating the samples into their constituent cell types is if you have more samples than cell types. Then the problem may become determined under a number of constraints (Venet et al., 2001):

$$M = GC,$$

where M is the matrix of measured values (rows correspond to genes and columns correspond to experiments), G is the expression profile of each cell type (rows correspond to genes and columns correspond to cell types), and C is the concentration matrix (rows correspond to cell types and columns correspond to measurements). Under a number of assumptions and constraints, it may be possible to find an optimal solution G and C from M (Venet et al., 2001).

4.8 SUMMARY

Whether you have intensities from a spotted array or Signal (use Li and Wong's weighted PM, if possible) from an Affymetrix chip, the following suggestions apply:

- The standard normalization with one factor to get the same average intensity in all chips is a good way to start, but it is not the best way to do it. Use signal dependent normalization if possible.

- Repeat each condition of the experiment (as a rule-of-thumb at least three times) and apply a statistical test for significance of observed differences. Apply the test on the normalized intensities (or expression indices). For spotted arrays with large variation between slides you can consider applying the statistical test on the fold change from each slide as well.

- Correct the statistical test for multiple testing (Bonferroni correction or similar).

Software for all the methods described in this chapter is available from www.bioconductor.org, which will be described in more detail in Chapter 14.

4.9 FURTHER READING

Vingron, M. (2001). Bioinformatics needs to adopt statistical thinking (Editorial). *Bioinformatics* 17:389–390.

Normalization

Bolstad, B. M., Irizarry, R. A., Astrand, M., Speed, T. P. (2003). A comparison of normalization methods for high density oligonucleotide array data based on variance and bias. *Bioinformatics* 19(2):185–193.

Dudoit, S., Yang, Y., Callow, M. J., and Speed, T. P. (2000). Statistical methods for identifying differentially expressed genes in replicated cDNA microarray experiments Technical report #578, August 2000.[8]

Goryachev, A. B., Macgregor, P. F., and Edwards, A. M. (2001). Unfolding of microarray data. *Journal of Computational Biology* 8:443–461.

Irizarry, R. A., Hobbs, B., Collin, F., Beazer-Barclay, Y. D., Antonellis, K. J., Scherf, U., and Speed, T. P. (2003) Exploration, normalization, and summaries of high density oligonucleotide array probe level data. *Biostatistics* 4(2):249–264.

Li, C., and Wong, W. H. (2001b). Model-based analysis of oligonucleotide arrays: Model validation, design issues and standard error application. *Genome Biology* 2:1–11.[9]

Schadt, E. E., Li, C., Su, C., and Wong, W. H. (2000). Analyzing high-density oligonucleotide gene expression array data. *J. Cell. BioChem.* 80:192–201.

Schuchhardt, J., Beule, D., Malik, A., Wolski, E., Eickhoff, H., Lehrach, H., and Herzel, H. (2000). Normalization strategies for cDNA microarrays. *Nucleic Acids Res.* 28:E47.

Workman, C., Jensen, L. J., Jarmer, H., Berka, R., Saxild, H. H., Gautier, L., Nielsen, C., Nielsen, H. B., Brunak, S, and Knudsen, S. (2002). A new non-linear normalization method for reducing variance between DNA microarray experiments. *Genome Biology* 3(9):0048.[10]

Zien, A., Aigner, T., Zimmer, R., and Lengauer, T. (2001). Centralization: A new method for the normalization of gene expression data. *Bioinformatics* 17(Suppl 1):S323–S331.

---

[8]Available at http://www.stat.berkeley.edu/tech-reports/index.html
[9]Software available at http://www.dchip.org
[10]Software available in affy package of Bioconductor http://www.bioconductor.org

Yang, Y. H., Dudoit, S., Luu, P., Lin, D. M., Peng, V., Ngai, J., and Speed, T. P. (2002). Normalization for cDNA microarray data: a robust composite method addressing single and multiple slide systematic variation. *Nucleic Acids Research* 30(4):e15.

Expression Index Calculation

Irizarry, R. A., Bolstad, B. M., Collin, F., Cope, L. M., Hobbs, B., and Speed, T. P. (2003) Summaries of Affymetrix GeneChip probe level data. *Nucleic Acids Research* 31(4):e15.

Lazaridis, E. N., Sinibaldi, D., Bloom, G., Mane, S., and Jove, R. (2002). A simple method to improve probe set estimates from oligonucleotide arrays. *Math. Biosci.* 176(1):53–58.

Lemon, W. J., Palatini, J. T., Krahe, R., and Wright, F. A. (2002). Theoretical and experimental comparisons of gene expression indexes for oligonucleotide arrays. *Bioinformatics* 18(11):1470–1476.

Li, C., and Wong, W. H. (2001a). Model-based analysis of oligonucleotide arrays: Expression index computation and outlier detection. *Proc. Natl. Acad. Sci. USA* 98:31–36.[11]

Zhang, L., Miles, M. F., and Aldape, K. D. (2003a). A model of molecular interactions on short oligonucleotide microarrays. *Nature Biotechnology* 21(7):818–821.

Zhang, L., Miles, M. F., and Aldape, K. D. (2003b) Corrigendum: A model of molecular interactions on short oligonucleotide microarrays. *Nature Biotechnology* 21(8):941.

Nonparametric Significance Tests Developed for Array Data

Efron, B., and Tibshirani, R. (2002). Empirical bayes methods and false discovery rates for microarrays. *Genet. Epidemiol.* 23(1):70–86.

Park, P. J., Pagano, M., and Bonetti, M. (2001). A nonparametric scoring algorithm for identifying informative genes from microarray Data. *Pacific Symposium on Biocomputing* 6:52–63.[12]

---

[11]Software available at http://www.dchip.org
[12]Manuscript available online at http://psb.stanford.edu

Student's t-test, ANOVA, and Wilcoxon/Mann-Whitney

Giles, P. J., and Kipling, D. (2003). Normality of oligonucleotide microarray data and implications for parametric statistical analyses. *Bioinformatics* 19: 2254–2262.

Kerr, M. K., Martin, M., and Churchill, G. A. (2000). Analysis of variance for gene expression microarray data. *J. Comput. Biol.* 7:819–837.

Kerr, M. K., and Churchill, G. A. (2001). Statistical design and the analysis of gene expression microarray data. *Genet Res.* 77:123–128. Review.

Montgomery, D. C., and Runger, G. C. (1999). *Applied Statistics and Probability for Engineers*. New York: Wiley.

Number of Replicates

Black, M. A., and Doerge, R. W. (2002). Calculation of the minimum number of replicate spots required for detection of significant gene expression fold change in microarray experiments. *Bioinformatics* 18(12):1609–1616.

Lee, M. L., and Whitmore, G. A. (2002). Power and sample size for DNA microarray studies. *Stat. Med.* 21(23):3543–3570.

Pan, W., Lin, J., and Le, C. (2002). How many replicates of arrays are required to detect gene expression changes in microarray experiments? A mixture model approach. *Genome Biology* 3(5):research0022.

Pavlidis, P., Li, Q., and Noble, W. S. (2003). The effect of replication on gene expression microarray experiments. *Bioinformatics* 19(13):1620–1627.

Piper, M. D., Daran-Lapujade, P., Bro, C., Regenberg, B., Knudsen, S., Nielsen, J., Pronk, J. T. (2002). Reproducibility of oligonucleotide microarray transcriptome analyses. An interlaboratory comparison using chemostat cultures of Saccharomyces cerevisiae. *J. Biol. Chem.* 277(40):37001–37008.

Wahde, M., Klus, G. T., Bittner, M. L., Chen, Y., Szallasi, Z. (2002). Assessing the significance of consistently mis-regulated genes in cancer associated gene expression matrices. *Bioinformatics* 18(3):389–394.

Correction for Multiple Testing

Bender, R., and Lange, S. (2001). Adjusting for multiple testing—when and how? *Journal of Clinical Epidemiology* 54:343–349.

Benjamini, Y, and Hochberg, Y. (1995). Controlling the false discovery rate: A practical and powerful approach to multiple testing. *J. R. Statist. Soc. B* 57(1):289–300.

Dudoit, S., Yang, Y., Callow, M. J., and Speed, T. P. (2000). Statistical methods for identifying differentially expressed genes in replicated cDNA microarray experiments. Technical report #578, August 2000.[13]

Reiner, A., Yekutieli, D., and Benjamini, Y. (2003). Identifying differentially expressed genes using false discovery rate controlling procedures. *Bioinformatics* 19(3):368–375.

Variance Stabilization

Baldi, P., and Long, A. D. (2001). A Bayesian framework for the analysis of microarray expression data: Regularized t-test and statistical inferences of gene changes. *Bioinformatics* 17:509–519.[14]

Huber, W., Von Heydebreck, A., Sultmann, H., Poustka, A., Vingron, M. (2002). Variance stabilization applied to microarray data calibration and to the quantification of differential expression. *Bioinformatics* 18 Suppl 1:S96–S104.

Lonnstedt, I. and Speed, T. (2002). Replicated Microarray Data. *Statistica Sinica* 12:31–46.

Other Significance Tests Developed for Array Data

Baggerly, K. A., Coombes, K. R., Hess, K. R., Stivers, D. N, Abruzzo, L. V., and Zhang, W. (2001). Identifying differentially expressed genes in cDNA microarray experiments. *Journal of Computational Biology* 8(6):639–659.

Ideker, T., Thorsson, V., Siegel, A. F., and Hood, L. (2000). Testing for differentially-expressed genes by maximum-likelihood analysis of microarray data. *Journal of Computational Biology* 7:805–817.

Newton, M. A., Kendziorski, C. M., Richmond, C. S., Blattner, F. R., and Tsui, K. W. (2001). On differential variability of expression ratios: Improving statistical inference about gene expression changes from microarray data. *Journal of Computational Biology* 8:37–52.

Rocke, D. M., and Durbin, B. (2001). A model for measurement error for gene expression arrays. *Journal of Computational Biology* 8(6):557–569.

---

[13] Available at http://www.stat.berkeley.edu/tech-reports/index.html
[14] Accompanying web page at http://visitor.ics.uci.edu/genex/cybert/

Theilhaber, J., Bushnell, S., Jackson, A., and Fuchs, R. (2001). Bayesian estimation of fold-changes in the analysis of gene expression: the PFOLD algorithm. *Journal of Computational Biology* 8(6):585–614.

Thomas, J. G., Olson, J. M., Tapscott, S. J., and Zhao, L. P. (2001). An efficient and robust statistical modeling approach to discover differentially expressed genes using genomic expression profiles. *Genome Res.* 11:1227–1236.

Tusher, V. G., Tibshirani, R., and Chu, G. (2001). Significance analysis of microarrays applied to the ionizing radiation response. *Proc. Natl. Acad. Sci. USA* 98:5119–5121.[15]

Zhao, L. P., Prentice, R., and Breeden, L. (2001). Statistical modeling of large microarray data sets to identify stimulus-response profiles. *Proc. Natl. Acad. Sci. USA* 98:5631–5636.

Wolfinger, R. D., Gibson, G., Wolfinger, E. D., Bennett, L., Hamadeh, H., Bushel, P., Afshari, C., and Paules, R. S. (2001). Assessing gene significance from cDNA microarray expression data via mixed models. *Journal of Computational Biology* 8(6):625–637.

Mixed Cell Populations

Lu, P., Nakorchevskiy, A., and Marcotte, E. M. (2003). Expression deconvolution: a reinterpretation of DNA microarray data reveals dynamic changes in cell populations. *Proc. Natl. Acad. Sci. USA* 100(18):10370–10375.

Venet, D., Pecasse, F., Maenhaut, C., and Bersini, H. (2001). Separation of samples into their constituents using gene expression data. *Bioinformatics* 17(Suppl 1):S279–S287.

---

[15] Software available for download at http://www-stat.stanford.edu/~tibs/SAM/index.html

5

Visualization by Reduction of Dimensionality

The data from expression arrays are of high dimensionality. If you have measured 6000 genes in 15 patients, the data constitute a matrix of 15 by 6000. It is impossible to discern any trends by visual inspection of such a matrix. It is necessary to reduce the dimensionality of this matrix to allow visual analysis. Since visual analysis is traditionally performed in two dimensions, in a coordinate system of x and y, many methods allow reduction of a matrix of any dimensionality to only two dimensions. These methods include principal component analysis, correspondence analysis, multidimensional scaling, and cluster analysis.

5.1 PRINCIPAL COMPONENT ANALYSIS

If we want to display the data in just two dimensions, we want to capture as much of the variation in the data as possible in just these two dimensions. Principal component analysis (PCA) has been developed for this purpose. Imagine 6000 genes as points in a 15-dimensional hyperspace, each dimension corresponding to expression in one of 15 patients. You will see a cloud of 6000 points in hyperspace. But the cloud is not hyperspherical. There will be one direction in which the cloud will be more irregular or extended. (Figure 5.1 illustrates this with only a few points in three dimensions.) This is the direction of the first principal component. This direction will not necessarily coincide with one of the patient axes. Rather, it will have projections of several, or all, patient axes on it. Next, we look for a direction that is orthogonal to the first

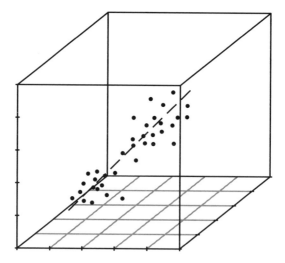

Fig. 5.1 A cloud of points in three-dimensional space. The cloud is not regular. It extends more in one direction than in all other directions. This direction is the first principal component (dashed line).

principal component, and captures the maximum amount of variation left in the data. This is the second principal component. We can now plot all 6000 genes in these two dimensions. We have reduced the dimensionality from 15 to 2, while trying to capture as much variation in the data as possible. The two principal components have been constructed as sums of the individual patient axes.

What will this analysis tell you? Perhaps nothing; it depends on whether there is a trend in your data that is discernible in two dimensions. Other relationships can be visualized with cluster analysis, which will be described in Chapter 6.

Instead of reducing the patient dimensions we can reduce the gene dimensions. Why not throw out all those genes that show no variation? We can achieve this by performing a principal component analysis of the genes. We are now imagining our data as 15 points in a space of 6000 dimensions, where each dimension records the expression level of one gene. Some dimensions contribute more to the variation between patients than other dimensions. The first principal component is the axis that captures most variation between patients. A number of genes have a projection on this axis, and the principal component method can tell you how much each gene contributes to the axis. The genes that contribute the most show the most variation between patients. Can this method be used for selecting diagnostic genes? Yes, but it is not necessarily the best method, because variation, as we have seen in Section 4.6, can be due to noise as well as true difference in expression. Besides, how

Table 5.1   Expression readings of four genes in six patients.

Gene	Patient					
	N_1	N_2	A_1	A_2	B_1	B_2
a	90	110	190	210	290	310
b	190	210	390	410	590	610
c	90	110	110	90	120	80
d	200	100	400	90	600	200

do we know how many genes to pick? The t-test and ANOVA, mentioned in Section 4.6, are more suited to this task. They will, based on a replication of measurement in each patient class, tell you which genes vary between patient classes and give you the probability of false positives at the cutoff you choose.

So a more useful application of principal component analysis would be to visualize genes that have been found by a t-test or ANOVA to be significantly regulated. This visualization may give you ideas for further analysis of the data.

5.2   EXAMPLE 1: PCA ON SMALL DATA MATRIX

Let us look at a simple example to visualize the problem. We have the data matrix shown in Table 5.1.

It consists of four genes measured in six patients. If we perform a principal component analysis on these data (the details of the computation are shown in Section 14.5), we get the biplot shown in Figure 5.2. A biplot is a plot designed to visualize both points and axes simultaneously. Here we have plotted the four genes as points in two dimensions, the first two principal components. It can be seen that genes a, c, and b differ a lot in the first dimension (they vary from about -400 to $+400$), while they differ little in the second dimension. Gene d, however, is separated from the other genes in the second dimension (it has a value of about -400 in the second dimension).

Indicated as arrows are the projections of the six patient axes on the two first principal components. Start with Patient B_1. This patient has a large projection (about 0.5) on the first principal component, and a smaller projection on the second principal component (about -0.3). The lengths of the patient vectors indicate how much they contribute to each axis and their directions indicate in which way they contribute. The first principal component consists mainly of Patient category B, where expression differs most. Going back to the genes, it can be seen that they are ranked according to average expression level in the B patients along this first principal component: genes c, a, d and b.

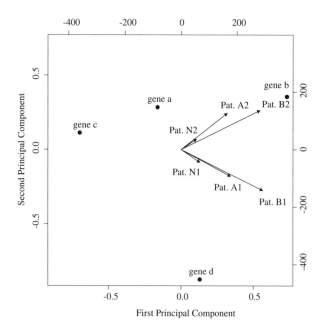

Fig. 5.2 Principal Component Analysis of data shown in Table 5.1. Note that patients and genes use different coordinate systems in this plot. Hence the different scales on the axes.

The second principal component divides genes into those that are higher in Patient B_2 than in Patient B_1 (gene c, a, and b), and gene d, which is lower in Patient B_2 than in Patient B_1. On the vector projections of the patient axes on this component it can be seen that they have been divided into those with subcategory 1 (Patients N_1, A_1, B_1), which all have a positive projection, and those with subcategory 2 (Patients N_2, A_2, B_2), which all have a negative projection. So the second principal component simply compares expression in subcategories.

We can also do the principal component analysis on the reverse (*transposed*) matrix (transposition means to swap rows with columns). Figure 5.3 shows a biplot of patients along principal components that consist of those genes that vary most between patients. First, it can be seen that there has been some grouping of patients into categories. Categories can be separated by two parallel lines. By looking at the projection of the gene vectors we can see that gene b and gene d, those that vary most, contribute most to the two axes. Now, if we wanted to use this principal component analysis to select genes that are diagnostic for the three categories, we might be tempted to select gene b and gene d because they contribute most to the first principal component. This would be a mistake, however, because gene d just shows high variance that is not correlated to category at all. The ANOVA, described

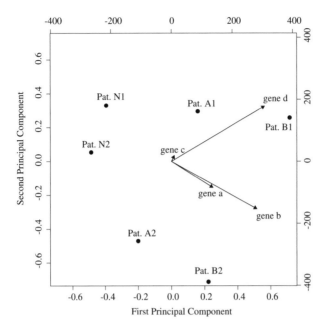

Fig. 5.3   Principal Component Analysis of transposed data of Table 5.1.

in Section 4.6, would have told us that gene b and gene a are the right genes to pick as diagnostic genes for the disease.

5.3   EXAMPLE 2: PCA ON REAL DATA

Figure 5.4 shows results of a PCA on real data. The R package was used on the HIV data (Section 1.5) as described in Section 14.5. The projections of the 7 experiments (4 controls (C) and 3 HIV (H)) on the principal components are shown as vectors in this biplot. The first principal component captures overall differences in expression level among genes—it separates them into those with negative expression (AvgDiff) and those with high expression. The second principal component separates the genes into those expressed higher in HIV than in the controls and those expressed higher in the controls than in HIV. What is ignored in this separation, however, is the variance between replicates in each group. The t-test (Section 4.6) makes a better selection of differentially expressed genes because it takes into account variance between replicates.

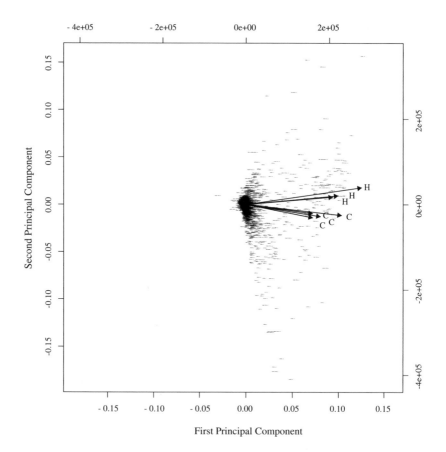

Fig. 5.4 Principal component analysis on real data from HIV experiment. All genes are plotted along the first two principal components. Genes are indicated by their name, but there are too many and the font is too small to be legible in this plot. The projections of the 7 experiments (4 controls (C) and 3 HIV (H)) on the principal components are shown as vectors in this biplot.

5.4 SUMMARY

Principal component analysis is a way to reduce your multidimensional data to a single $x - y$ graph. You may be able to spot important trends in your data from this one graph alone. If replicates are available it is best to perform PCA on data that has already been filtered for significance.

5.5 FURTHER READING

Singular Value Decomposition

Alter, O., Brown, P. O., and Botstein, D. (2000). Singular value decomposition for genome-wide expression data processing and modeling. *Proc. Natl. Acad. Sci. USA* 97:10101–10106.

Holter, N. S., Mitra, M., Maritan, A., Cieplak, M., Banavar, J. R., and Fedoroff, N.V. (2000). Fundamental patterns underlying gene expression profiles: Simplicity from complexity. *Proc. Natl. Acad. Sci. USA* 97:8409–8414.

Wall, M. E., Dyck, P. A., and Brettin, T. S. (2001). SVDMAN—singular value decomposition analysis of microarray data. *Bioinformatics* 17:566–568.

Principal Component Analysis

Dysvik, B, and Jonassen, I. (2001). J-Express: Exploring gene expression data using Java. *Bioinformatics* 17:369–370.[1]

Raychaudhuri, S., Stuart, J. M., and Altman, R. B. (2000). Principal components analysis to summarize microarray experiments: Application to sporulation time series. *Pac. Symp. Biocomput.* 2000:455–466.[2]

Xia, X, and Xie, Z. (2001). AMADA: Analysis of microarray data. *Bioinformatics* 17:569–570.

Xiong, M., Jin, L., Li, W., and Boerwinkle, E. (2000). Computational methods for gene expression-based tumor classification. *Biotechniques* 29:1264–1268.

Correspondence Analysis

---

[1] Software available at http://www.ii.uib.no/~bjarted/jexpress/
[2] Available online at http://psb.stanford.edu

Fellenberg, K., Hauser, N. C., Brors, B., Neutzner, A., Hoheisel, J. D., and Vingron, M. (2001). Correspondence analysis applied to microarray data. *Proc. Natl. Acad. Sci. USA* 98:10781–10786.

Gene Shaving uses PCA to Select Genes with Maximum Variance

Hastie, T., Tibshirani, R., Eisen, M. B., Alizadeh, A., Levy, R., Staudt, L., Chan, W. C., Botstein, D., and Brown, P. (2000). Gene shaving as a method for identifying distinct sets of genes with similar expression patterns. *Genome Biol.* 1:RESEARCH0003.1–21

6

Cluster Analysis

If you have just one experiment and a control, your first data analysis will limit itself to a list of regulated genes ranked by the magnitude of up- and downregulation, or ranked by the significance of regulation determined in a *t*-test.

Once you have more experiments—measuring the same genes under different conditions, in different mutants, in different patients, or at different time points during an experiment—it makes sense to group the significantly changed genes into clusters that behave similarly over the different conditions.

6.1 HIERARCHICAL CLUSTERING

Think of each gene as a vector of N numbers, where N is the number of experiments or patients. Then you can plot each gene as a point in N-dimensional hyperspace. You can then calculate the distance between two genes as the Euclidean distance between their respective data points (as the root of the sum of the squared distances in each dimension).

This can be visualized using a modified version of the small example data set applied in previous chapters (Table 6.1). The measured expression level of the five genes can be plotted in just two of the patients using a standard $x - y$ coordinate system (Figure 6.1 left).

You can calculate the distance between all genes (producing a *distance matrix*), and then it makes sense to group those genes together that are closest to each other in space. The two genes that are closest to each other, b and d,

Table 6.1   Expression readings of five genes in two patients.

Gene	Patient	
	N_1	A_1
a	90	190
b	190	390
c	90	110
d	200	400
e	150	200

form the first cluster (Figure 6.1 left). Genes a and c are separated by a larger distance, and they form a cluster as well (Figure 6.1 left). If the separation between a gene and a cluster comes within the distance as you increase it, you add that gene to the cluster. Gene e is added to the cluster formed by a and c. How do you calculate the distance between a point (gene) and a cluster? You can calculate the distance to the *nearest neighbor* in the cluster (gene a), but it is better to calculate the distance to the point that is in the middle of the existing members of the cluster (centroid, similar to UPGMA or average linkage method).

When you have increased the distance to a level where all genes fall within that distance, you are finished with the clustering and can connect the final

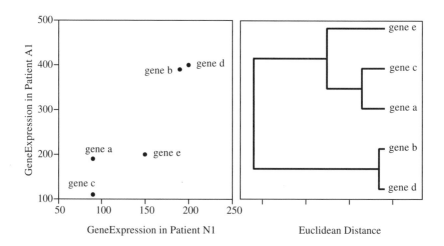

Fig. 6.1   Hierarchical clustering of genes based on their Euclidean distance visualized by a rooted tree. Note that it is possible to reorder the leaves of the tree by flipping the branches at any node without changing the information in the tree.

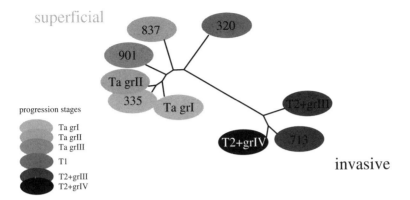

Fig. 6.2 Hierarchical clustering of bladder cancer patients using an unrooted tree. The clustering was based on expression measurements from a DNA array hybridized with mRNA extracted from a biopsy. Numbers refer to patients and the severity of the disease is indicated by a color code. (Christopher Workman, based on data published in Thykjaer et al. (2001)). (See color plate.)

clusters. You have now performed a *hierarchical agglomerative clustering*. There are computer algorithms available for doing this (see Section 14.4).

A real example is shown in Figure 6.2, where bladder cancer patients were clustered based on Affymetrix GeneChip expression measurements from a bladder biopsy. It is seen in the figure that the clustering groups superficial tumors together and groups invasive tumors together.

Hierarchical clustering only fails when you have a large number of genes (several thousand). Calculating the distances between all of them becomes time consuming. Removing genes that show no significant change in any experiment is one way to reduce the problem. Another way is to use a faster algorithm, like *K*-means clustering.

6.2 *K*-MEANS CLUSTERING

In *K*-means clustering, you skip the calculation of distances between all genes. You decide on the number of clusters you want to divide the genes into, and the computer then randomly assigns each gene to one of the K clusters. Now it will be comparatively fast to calculate the distance between each gene and the center of each cluster (*centroid*). If a gene is actually closer to the center of another cluster than the one it is currently assigned to, it is reassigned to the closer cluster. After assigning all genes to the closest cluster, the centroids are recalculated. After a number of iterations of this, the cluster centroids will no longer change, and the algorithm stops. This is a very fast algorithm, but it will give you only the number of clusters you asked

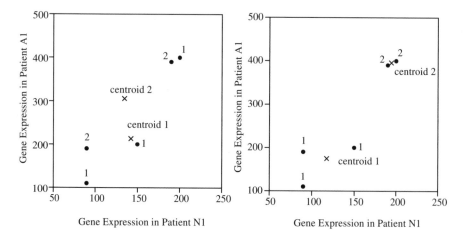

Fig. 6.3 *K*-means clustering of genes based on their Euclidean distance. First, genes are randomly assigned to one of the two clusters in *K*: 1 or 2 (Left). The centroids of each cluster are calculated. Genes are then reassigned to another cluster if they are closer to the centroid of that cluster (Right). In this simple example, the final solution is obtained after just one iteration (Right).

for and not show their relation to each other as a full hierarchical clustering will do. In practice, *K*-means is useful if you try different values of *K*.

If you try the *K*-means clustering on the expression data used for hierarchical clustering shown in Figure 6.1, with $K = 2$, the algorithm may find the solution in just one iteration (Figure 6.3).

6.3 SELF-ORGANIZING MAPS

There are other methods for clustering, but hierarchical and *K*-means cover most needs. One method that is available in a number of clustering software packages is self-organizing maps (SOM) (Kohonen, 1995). SOM is similar to *K*-means, but clusters are ordered on a low-dimensional structure, such as a grid. The advantage over *K*-means is that neighboring clusters in this grid are more related than clusters that are not neighbors. So it results in an ordering of clusters that is not performed in *K*-means.

Figure 6.4 shows how a SOM clustering could fit a two-by-two grid on the data of our example (the cluster centers are at each corner of the grid). That would result in four clusters, three of them with one member only and one cluster with two members.

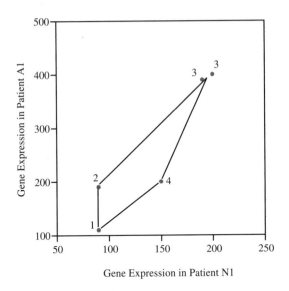

Gene Expression in Patient N1

Fig. 6.4   SOM clustering of genes into a two-by-two grid, resulting in four clusters.

6.4 DISTANCE MEASURES

In addition to calculating the Euclidean distance, there are a number of other ways to calculate distance between two genes. When these are combined with different ways of normalizing your data, the choice of normalization and distance measure can become rather confusing. Here I will attempt to show how the different distance measures relate to each other and what effect, if any, normalization of the data has. Finally I will suggest a good choice of distance measure for expression data.

The Euclidean distance between two points a and b in N-dimensional space is defined as

$$\sqrt{\sum_{i=1}^{N} (a_i - b_i)^2},$$

where i is the index that loops over the dimensions of N, and the Σ sign indicates that the squared distances in each dimension should be summed before taking the square root of those sums. Figure 6.5 shows the Euclidean distance between two points in two-dimensional space.

Instead of calculating the Euclidean distance, you can also calculate the angle between the vectors that are formed between the data point of the gene and the center of the coordinate system. For gene expression, *vector angle* (Fig. 6.5) often performs better because the trend of a regulation response is

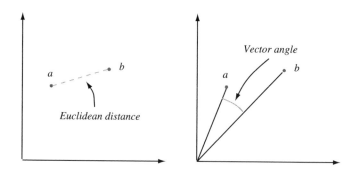

Fig. 6.5 Euclidean distance and vector angle between points a and b in two-dimensional space.

more important than its magnitude. Vector angle α between points a and b in N-dimensional space is calculated as

$$\cos \alpha = \frac{\sum_{i=1}^{N} a_i b_i}{\sqrt{\sum_{i=1}^{N} a_i^2}\sqrt{\sum_{i=1}^{N} b_i^2}}.$$

Finally, a widely used distance metric is the Pearson correlation coefficient:

$$\mathrm{PearsonCC} = \frac{\sum_{i=1}^{N}(a_i - \bar{a})(b_i - \bar{b})}{\sqrt{\sum_{i=1}^{N}(a_i - \bar{a})^2}\sqrt{\sum_{i=1}^{N}(b_i - \bar{b})^2}}.$$

You can see that the only difference between vector angle and Pearson correlation is that the means ($\bar{a}$ and $\bar{b}$) have been subtracted before calculating the Pearson correlation. So taking the vector angle of a means-normalized data set (each gene has been centered around its mean expression value over all conditions) is the same as taking the Pearson correlation.

An example will illustrate this point. Let us consider two genes, a and b, that have the expression levels $a = (1, 2, 3, 4)$ and $b = (2, 4, 6, 8)$ in four experiments. They both show an increasing expression over the four experiments, but the magnitude of response differs. The Euclidean distance is 5.48, while the vector angle distance ($1 - \cos \alpha$) is zero and the Pearson distance (1-Pearson CC) is zero. I would say that because the two genes show exactly the same trend in the four experiments, the vector angle and Pearson distance make more sense in a biological context than the Euclidean distance.

Table 6.2 Expression readings of four genes in six patients

Gene	Patient					
	N_1	N_2	A_1	A_2	B_1	B_2
a	90	110	190	210	290	310
b	190	210	390	410	590	610
c	90	110	110	90	120	80
d	200	100	400	90	600	200

Table 6.3 Euclidean distance matrix between four genes.

Gene	Gene			
	a	b	c	d
a	0.00	5.29	3.20	4.23
b	5.29	0.00	8.38	5.32
c	3.20	8.38	0.00	5.84
d	4.23	5.32	5.84	0.00

Table 6.4 Vector angle distance matrix between four genes.

Gene	Gene			
	a	b	c	d
a	0.00	0.02	0.42	0.52
b	0.02	0.00	0.41	0.50
c	0.42	0.41	0.00	0.51
d	0.52	0.50	0.51	0.00

6.4.1 Example: Comparison of Distance Measures

Let us try the different distance measures on our little example of four genes from six patients (Table 6.2). We can calculate the Euclidean distances (see Section 14.4 for details on how to do this) between the four genes. The pairwise distances between all genes can be shown in a *distance matrix* (Table 6.3) where the distance between gene a and a is zero, so the pairwise identities form a diagonal of zeros through the matrix. The triangle above the diagonal is a mirror image of the triangle below the diagonal because the distance between genes a and b is the same as the distance between genes b and a.

This distance matrix is best visualized by clustering as shown in Figure 6.6, where it is compared with clustering based on vector angle distance (Table 6.4) and a tree based on Pearson correlation distances (Table 6.5).

Table 6.5   Pearson distance matrix between four genes.

Gene	Gene			
	a	*b*	*c*	*d*
a	0.00	0.06	1.45	1.03
b	0.06	0.00	1.43	0.98
c	1.45	1.43	0.00	0.83
d	1.03	0.98	0.83	0.00

The clustering in Figure 6.6 is nothing but a two-dimensional visualization of the four-by-four distance matrix. What does it tell us? It tells us that vector angle distance is the best way to represent gene expression responses. Genes a, b, and d all have increasing expression over the three patient categories, only the magnitude of the response and the error between replicates differs. The vector angle clustering has captured this trend perfectly, grouping a and b close together and d nearby. Euclidean distance has completely missed this picture, focusing only on absolute expression values, and placed genes a and b furthest apart. Pearson correlation distance has done a pretty good job, capturing the close biological proximity of genes a and b, but it has normalized the data too heavily and placed gene d closest to gene c, which shows no trend in the disease at all.

It is also possible to cluster in the other dimension, clustering patients instead of genes. Instead of looking for genes which show a similar transcriptional response to the progression of a disease, we are looking for patients that have the same *profile* of expressed genes. If two patients have exactly the same stage of a disease we hope that this will be reflected in identical expression of a number of key genes. Thus, we are not interested in the genes that are not expressed in any patient, are unchanged between patients, or show a high error. So it makes sense to remove those genes before clustering

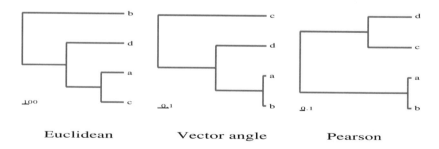

Euclidean   Vector angle   Pearson

Fig. 6.6   Hierarchical clustering of distances (with three different distance measures) between genes in the example.

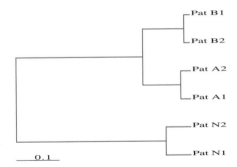

Pat B1
Pat B2
Pat A2
Pat A1
Pat N2
Pat N1
0.1

Fig. 6.7 Hierarchical clustering of vector angle distances between patients in the example.

patients. But that requires applying a t-test or ANOVA on the data and in order to do so you have to put the data into the categories you presume the patients fall into. That could obscure trends in the data that you have not yet considered. In practice, you could try clustering both on the full data and on data cleaned by t-test, ANOVA, or another method.

The data in our little example contains too much noise from gene c to cluster on the complete set of genes. If an ANOVA is run on the three patient categories (Section 4.6) that only leaves genes a and b, and the hierarchical vector angle distance clustering based on those two genes can be seen in Figure 6.7. But remember that these two genes were selected explicitly for the purpose of separating the three patient categories.

6.5 TIME-SERIES ANALYSIS

When your data consist of samples from different time points in an experiment, this presents a unique situation for analysis. There are two fundamentally different ways of approaching the analysis. Either you can take replicate samples from each time point and use statistical methods such as t-test or ANOVA to identify genes that are expressed differentially over the time series. This approach does not rely on any assumptions about the spacing between time points, or the behavior of gene expression over time.

Another way of approaching the analysis is to assume that there is some relationship between the time points. For example, you can assume that there is a linear relationship between the samples, so that genes increase or decrease in expression in a linear manner over time. In that case you can use linear modeling as a statistical analysis tool.

Another possible relationship is a sine wave for cyclical phenomena. The sine wave has been used to analyse cell cycle experiments in yeast (Spellman et al., 1998).

If you have no prior expectation on the response in your data, clustering may be the most powerful way of discovering temporal relationships.

6.6 GENE NORMALIZATION

The difference between vector angle distance and Pearson correlation comes down to a means normalization. There are two other common ways of normalizing the expression level of a gene—length normalization and SD normalization:

- Means: Calculate mean and subtract from all numbers.

- Length: Calculate length of gene vector and divide all numbers by that length.

- SD: Calculate standard deviation and divide all numbers by it.

For each of these normalizations it is important to realize that it is performed on each gene in isolation; the information from other genes is not taken into account. Before you perform any of these normalizations, it is important that you answer this question: Why do you want to normalize the data in that way? Remember, you have already normalized the chips, so expression readings should be comparable. In general, normalization affects Euclidean distances to a large extent, it affects vector angles to a much smaller extent, and it hardly ever affects Pearson distances because the latter metric is normalized already. My suggestion for biological data is to use vector angle distance on non-normalized expression data for gene clustering.

6.7 VISUALIZATION OF CLUSTERS

Clusters are traditionally visualized with trees (Figures 6.7, 6.6, 6.2, and 6.1). Note that information is lost in going from a full distance matrix to a tree visualization of it. Different trees can represent the same distance matrix.

In DNA chip analysis it has also become common to visualize the gene vectors by representing the expression level or fold change in each experiment with a color-coded matrix. Figure 6.8 shows such a visualization of gene expression data using both a tree and a color matrix using the ClustArray software (Section 14.4).

6.7.1 Example: Visualization of Gene Clusters in Bladder Cancer

Figure 6.8 is a visualization of the most important genes (selected by their co-variance to the progression of the disease) in DNA microarray measurements in bladder cancer patients.

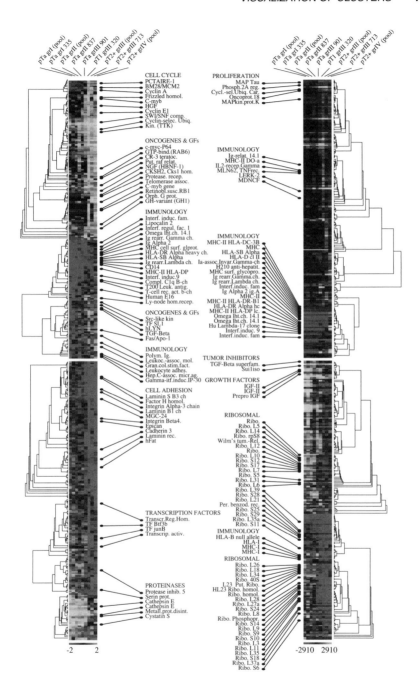

Fig. 6.8 Hierarchical clustering of genes (rows) expressed in bladder cancers (columns). Yellow fields show up-regulation of genes (absolute difference in right panel, logfold change in left panel), blue fields show down-regulation of genes. (Figure by Christopher Workman using ClustArray software on data from Thykjaer et al., (2001) and postprocessing with Adobe Illustrator). (See color plate.)

6.8 SUMMARY

Cluster analysis groups genes according to how they behave in experiments. For gene expression, measuring similarity of gene expression using the vector angle between expression profiles of two genes makes most sense. Normalization of your data matrix (of genes versus experiments) can be performed in either of two dimensions. If you normalize columns you normalize the total expression level of each experiment. A normalization of experiments to yield the same sum of all genes is referred to in this book as scaling and is described in Section 4.1. Such a normalization is essential before comparison of experiments, but a multifactor scaling with a spline or a polinomial is even better.

Normalization of genes in the other dimension may distort the scaling of experiments that you have performed (if you sum the expression of all genes in an experiment after a gene normalization, it will no longer add up to the same number). Also, normalization of genes before calculating vector angle is usually not necessary. Therefore, Pearson correlation is not quite as good a measure of similarity as vector angle.

6.9 FURTHER READING

Thykjaer, T., Workman, C., Kruhøffer, M., Demtröder, K., Wolf, H., Andersen, L. D., Frederiksen, C. M., Knudsen, S., and Ørntoft, T. F. (2001). Identification of gene expression patterns in superficial and invasive human bladder cancer. *Cancer Research* 61:2492–2499.

Clustering Methods and Cluster Reliability

Ben-Hur, A., Elisseeff, A., and Guyon, I. (2002). A Stability Based Method for Discovering Structure in Clustered Data. *Pacific Symposium on Biocomputing* 2002:6–17.[1]

De Smet, F., Mathys, J., Marchal, K., Thijs, G., De Moor, B., and Moreau, Y. (2002). Adaptive quality-based clustering of gene expression profiles. *Bioinformatics* 18(5):735–746.

Getz, G., Levine, E., and Domany, E. (2000). Coupled two-way clustering analysis of gene microarray data. *Proc. Natl. Acad. Sci. USA* 97:12079–12084.

Hastie, T., Tibshirani, R., Eisen, M. B., Alizadeh, A., Levy, R., Staudt, L., Chan, W. C., Botstein, D., and Brown, P. (2000). Gene shaving as a

---

[1] Available online at http://psb.stanford.edu

method for identifying distinct sets of genes with similar expression patterns. *Genome Biol.* 1:RESEARCH0003.1–21.

Herrero, J., Valencia, A., and Dopazo, J. (2001). A hierarchical unsupervised growing neural network for clustering gene expression patterns. *Bioinformatics* 17:126–136.

Kerr, M. K., and Churchill, G. A. (2001). Bootstrapping cluster analysis: Assessing the reliability of conclusions from microarray experiments. *Proc. Natl. Acad. Sci. USA* 98:8961–8965.

Kohonen, T. (1995). *Self-Organizing Maps.* Berlin: Springer.

Michaels, G. S., Carr, D. B., Askenazi, M., Fuhrman, S., Wen, X., and Somogyi, R. (1998). Cluster analysis and data visualization of large-scale gene expression data. *Pacific Symposium on Biocomputing* 3:42–53.[2]

Sasik, R., Hwa, T., Iranfar, N., and Loomis, W. F. (2001). Percolation clustering: A novel algorithm applied to the clustering of gene expression patterns in dictyostelium development. *Pacific Symposium on Biocomputing* 6:335–347.[3]

Spellman, P., Sherlock, G., Zhang, M., Lyer, V., Anders, K., Eisen, M., Brown, P., Botstein, D., and Futcher, B. (1998). Comprehensive identification of cell cycle-regulated genes of yeast *S. cerevisiae* by microarray hybridization. *Mol. Biol. Cell* 9:3273–3297.

Tamayo, P., Slonim, D., Mesirov, J., Zhu, Q., Kitareewan, S., Dmitrovsky, E., Lander, E. S., and Golub, T. R. (1999). Interpreting patterns of gene expression with self-organizing maps: Methods and application to hematopoietic differentiation. *Proc. Natl. Acad. Sci. USA* 96:2907–2912.

Tibshirani, R., Walther, G., Botstein, D., and Brown, P. (2000). Cluster validation by prediction strength. Technical report. Statistics Department, Stanford University.[4]

Xing, E. P., and Karp, R. M. (2001). CLIFF: Clustering of high-dimensional microarray data via iterative feature filtering using normalized cuts. *Bioinformatics* 17(Suppl 1):S306–S315.

Yeung, K. Y., Haynor, D. R., and Ruzzo, W. L. (2001). Validating clustering for gene expression data. *Bioinformatics* 17:309–318.

---

[2] Available online at http://psb.stanford.edu
[3] Available online at http://psb.stanford.edu
[4] Manuscript available at http://www-stat.stanford.edu/~tibs/research.html

Yeung, K. Y., Fraley, C., Murua, A., Raftery, A. E., and Ruzzo, W. L. (2001) Model-based clustering and data transformations for gene expression data. *Bioinformatics* 17:977–987.

7

Beyond Cluster Analysis

One day we may look back and understand how computation and experimentation with biological systems blurred the divide and allowed the 'great crossing' between the inanimate and the animate worlds.

—Ouzounis and Valencia, 2003

7.1 FUNCTION PREDICTION

Genes that appear in the same cluster have similar transcription response to different conditions. It is likely that this is caused by some commonality in function or role. If a cluster is populated by genes with known function—and that function is similar—you can infer the function of orphan genes in the same cluster.

Another valuable tool to assigning function, in particular for clusters where there are no genes with known function, is function prediction. Function prediction enters the scene where there is no sequence homology to proteins with known function. Instead, a number of properties and predicted features of the protein can be used to predict a likely function class (Jensen et al., 2002). It turns out that proteins with similar function also share some similarities in amino acid sequence length, posttranslational modification, cellular destination signal, and so on. Taken separately, each of these features is a weak predictor of function category. Taken together, a sufficiently large number of features can be used to make fairly accurate predictions of function class[1].

7.2   DISCOVERY OF REGULATORY ELEMENTS IN PROMOTER REGIONS

If a number of genes share a regulatory response to a number of stimuli it is reasonable to assume that they do so because they share a binding site for a transcription factor in their promoter.

The ClustArray web-based clustering software (Section 14.4) allows you to select the genes in a cluster and search their upstream promoter region for such common regulatory elements that can account for the similarity in transcription response. This works only for organisms where the promoter regions of the genes on the chip are known and included in a database. Currently, databases of promoter regions in human and *Saccharomyces cerevisiae* are available at the website.

Even when you have the promoter sequence, finding common regulatory elements is inherently complicated because of the degeneracy of such elements. You can use software such as *saco-patterns* (Jensen and Knudsen, 2000) to search for patterns that are fully conserved. An example of a fully conserved pattern is AGCTTAGG. Such a search is reasonably fast, simple, and deterministic, because it is possible to search for all possible patterns up to a given length. But it will miss all those patterns that are not conserved enough to be picked up by a single pattern such as AGCTTAGG. Transcription factor binding sites are typically *degenerate*; they tolerate some variation in sequence at some locations in the site. The problem is that there is an infinite number of possible degenerate sites. Still, software solutions have been developed for this problem. Degenerate patterns can be searched with software like *ann-spec* (Workman and Stormo, 2000), but it is sensitive to the choice of parameters, and it will not give the same result every time you run it. It uses a probabilistic Gibbs sampling approach to guess parameters for a weight matrix that describes the regulatory elements. Lawrence's Gibbs sampler (Neuwald and Lawrence, 1995; Lawrence et al., 1993), uses a similar strategy.

Running any of these methods to discover regulatory elements will lead you into an assessment of the significance of any discoveries. There are two good ways of assessing this. First, you can look for the occurrence of the discovered element in a background set, either in a set of promoters known not to contain the element, or in a set of all promoters in that organism, where you can assume that most promoters do not contain the element. Then you can compare the frequency of elements in the promoters in your positive set

---

[1] A web server is available at http://www.cbs.dtu.dk/services/ProtFun/

to the frequency of promoters in the background set and perform a statistical analysis (sampling without replacement) to calculate the probability that both sets have the same occurrence of the elements. If they do, then you have not found a biologically relevant element. Remember to correct for multiple testing (Bonferroni) before evaluating probabilities (see Section 4.6). Saco-patterns includes a statistical evaluation with Bonferroni correction.

Confirmation of a biologically relevant signal may also come from an observation that those genes in the background set that do contain the element actually are additional members of the pathway or function class of the positive set.

Finally, a way of assessing significance is to plot a histogram of positional preference of the signal relative to the transcription start site. Any obvious preference is a substantiation of biological significance, while an absence does not rule out significance.

If you perform any of the tests described above, be sure to perform them on promoter extracts of identical length to avoid artifacts of analysis.

7.2.1 Example 1: Discovery of Proteasomal Element

If you take all 6269 ORFs annotated in the GenBank file of *Saccharomyces cerevisiae* and extract 200 bp starting 300 bp upstream of the ORF, you cover most promoter regions in the organism pretty well. If you divide these 6269 promoter regions into those that have been annotated as related to the proteasome (31) and those that have not (6238), you have a positive set and a background set, respectively. If you run saco-patterns using these sets as positive and background, it finds the sequence GGTGGCAAA present in 25 of the positive set and 26 of the background set. That is such a vast over-representation that the probability that it is not significant—even after correction for multiple testing—is less than 10^{-10}. Of the 26 apparent false positives in the background set, two are proteases and three are genes related to ubiquitin, all of which could very well be coregulated with proteasomes (Jensen and Knudsen, 2000).

Note that in this example we did not use expression analysis such as *t*-test or clustering to generate a positive set of promoters. We used functional annotation. You can use any method to generate a positive set and then search for patterns overrepresented in that set.

7.2.2 Example 2: Rediscovery of Mlu Cell Cycle Box (MCB)

Using the yeast promoter regions from the previous example, but instead sorting them by the expression in one of the cell cycle experiments (Spellman et al., 1998), allows identification of patterns that are correlated with expression: Instead of dividing promoter regions into a positive and negative set, we look for patterns that are more frequent in up-regulated genes than in nonregulated

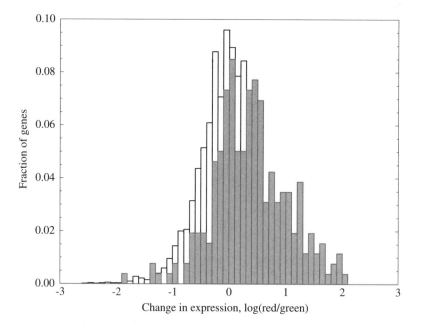

Fig. 7.1 Logfold distribution of all yeast genes (open bars) in a cell cycle experiment. Log-fold distribution of genes containing Mlu cell cycle box (shaded bars) in the same experiment. Drawing by Lars Juhl Jensen based on data from Jensen and Knudsen (2000).

genes or in down-regulated genes, or vice versa. There is a statistical test for this and it is called the Kolmogorov-Smirnov rank test (Jensen and Knudsen, 2000). Figure 7.1 shows the distribution of genes that contain such a pattern, the Mlu cell cycle Box (MCB). The well-known MCB pattern, ACGCGT, was discovered to be significant by saco-patterns testing all possible patterns up to length 8 in the cell cycle experiment.

7.3 SUMMARY

It is beyond the cluster analysis that the real data mining takes place: you can mine your data for promoter elements involved in the regulation that you observe, you can mine for novel functions of orphan proteins, you can mine for novel regulatory relationships between genes under study. In the future, these analyses should be combined to increase their power in detecting subtle relationships that may today be obscured by noise in your data.

Prepare Sample   Print Microarray

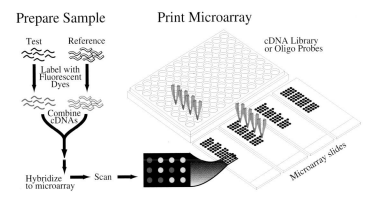

Fig. 1.5   The spotted array technology. A robot is used to transfer probes in solution from a microtiter plate to a glass slide where they are dried. Extracted mRNA from cells is converted to cDNA and labeled fluorescently. Sample is labeled red and control is labeled green. After mixing, they are hybridized to the probes on the glass slide. After washing away unhybridized material, the chip is scanned with a confocal laser and the image analyzed by computer.

Fig. 1.6   Spotted array containing more than 9000 features. Probes against each predicted open reading frame in *Bacillus subtilis* are spotted twice on the slide. Image shows color overlay after hybridization of sample and control and scanning. (Picture by Hanne Jarmer.)

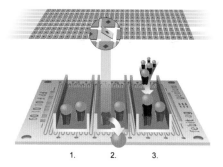

Fig. 1.8   Graphical illustration of the *in situ* synthesis of probes inside the Febit DNA processor.   Shown are three enlargements of a microchannel, each illustrating one step in the synthesis.   1: the situation before synthesis.   2: selected nucleotides are deprotected by controlling light illumination via a micromirror.   3: substrate is added to the microchannel and covalently attached to the deprotected positions.   (Copyright Febit AG. Used with permission.)

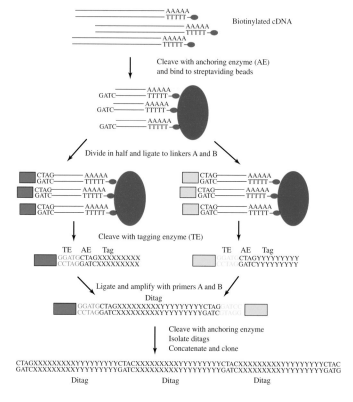

Fig. 1.9   Schematic overview of SAGE methods (based on Velculescu et al. 1995).

Fig. 1.10   cDNA microarray of genes affected by HIV infection.

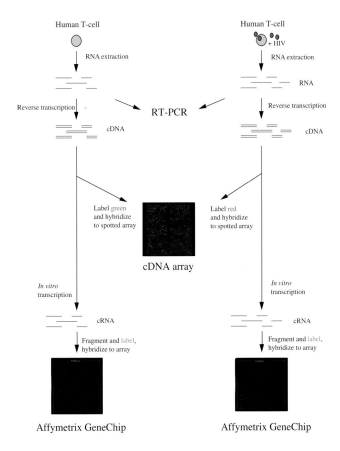

Fig. 1.14   Overview of methods for comparing mRNA populations in cells from two different conditions.

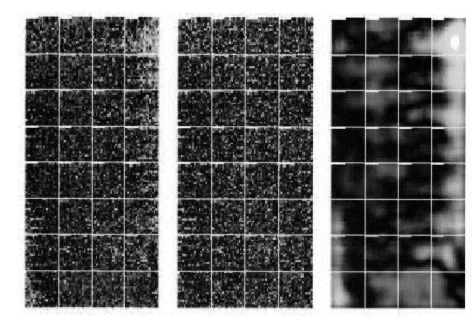

Fig. 3.4   Spatial effects on a spotted array. The blue-yellow color scale indicates fold change between the two channels. A spatial bias is visible (left). Gaussian smoothing captures the bias (right) which can the be removed by subtraction from the image (center). From Workman (2002).

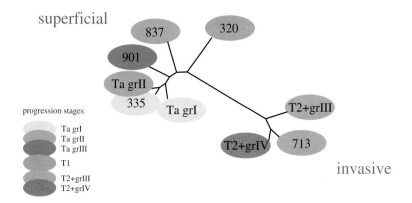

Fig. 6.2   Hierarchical clustering of bladder cancer patients. The clustering was based on expression measurements from a DNA array hybridized with mRNA extracted from a biopsy. (Christopher Workman, based on data published in Thykjaer et al. (2001)).

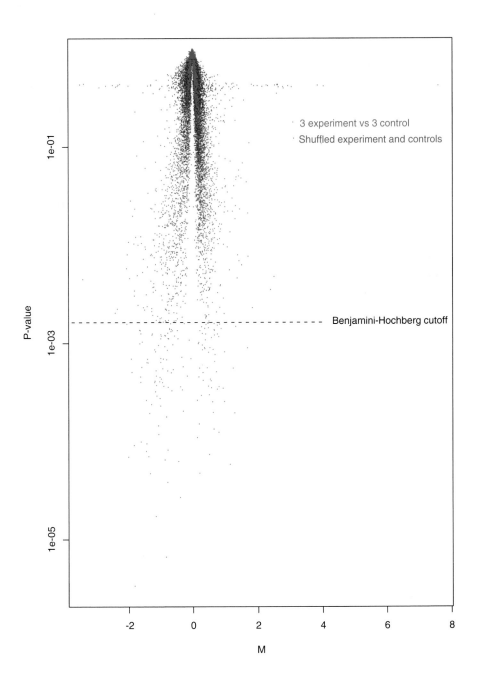

Fig. 4.4   Volcano plot showing relationship between *P*-value and logfold change (M) (Wolfinger, 2001). M is the average of three replicates of each condition. For comparison a random permutation of the data has been included (green points).

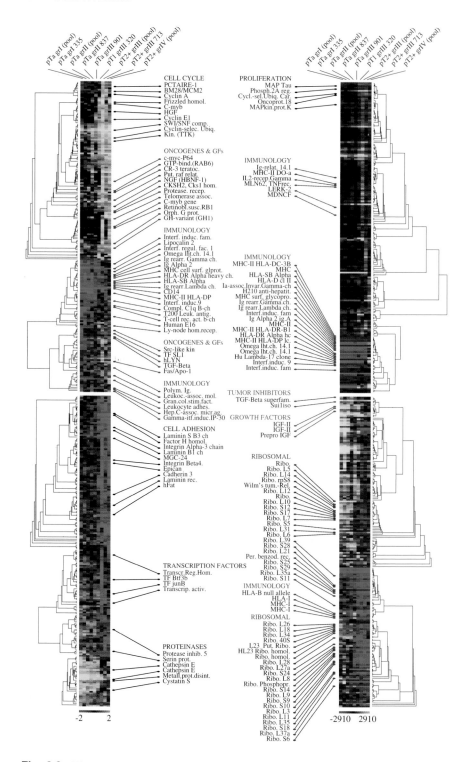

Fig. 6.8 Hierarchical clustering of genes (rows) expressed in bladder cancers (columns). Yellow fields show up-regulation of genes (absolute difference in right panel, logfold change in left panel), blue fields show down-regulation of genes. (Figure by Christopher Workman using ClustArray software on data from Thykjaer et al., (2001) and postprocessing with Adobe Illustrator).

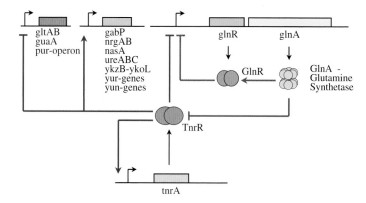

Fig. 9.2 Known regulatory network in *Bacillus subtilis*. Each line ending in a bar represents a deduced negative regulatory effect. Each line ending in an arrow represents a deduced positive regulatory effect. (Hanne Jarmer and Carsten Friis.)

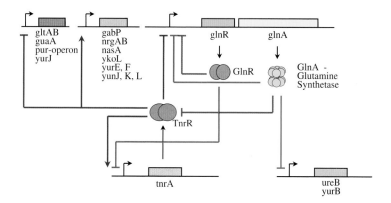

Fig. 9.3 Regulatory network reverse engineered from real steady-state data. Each line ending in a bar represents a deduced negative regulatory effect. Each line ending in an arrow represents a deduced positive regulatory effect. (Hanne Jarmer and Carsten Friis.)

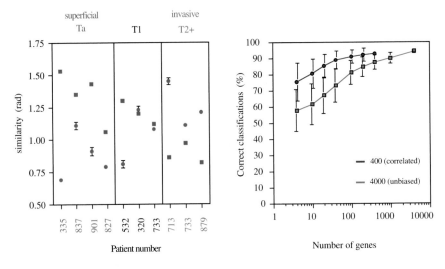

Fig. 10.3 Classifier of bladder cancers based on expression array. Left: Vector angle between patient and reference pool. Each three-digit number on the bottom refers to a patient. The angle between that patient and the two reference pools (squares, Ta pool; circles T2 pool) is indicated. The angle is always smallest (the similarity is greatest) to the pool with the same type of cancer. The intermediate type, T1, for which there is no reference pool, is sometimes more similar to one reference, sometimes more similar to another reference. Error bars have been added to show variation due to choice of reference pool, of which several were available. Right: the performance of the classifier as a function of the number of genes used for classification. Top curve: genes chosen among those 400 genes maximally covarying with the disease. Bottom curve: genes chosen at random from all 4000 genes detected as present in at least one patient.

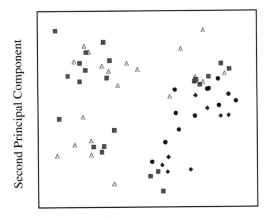

First Principal Component

Fig. 10.4 Principal component analysis of 63 small, round blue cell tumors.

7.4 FURTHER READING

Jensen, L. J., Gupta, R., Blom, N., Devos, D., Tamames, J., Kesmir, C., Nielsen, H., Stærfeldt, H. H., Rapacki, K., Workman, C., Andersen, C. A. F., Knudsen, S., Krogh, A., Valencia, A., and Brunak., S. (2002). Ab initio prediction of human orphan protein function from post-translational modifications and localization features. *Journal of Molecular Biology* 319:1257–1265.

Ouzounis, C. A., and Valencia, A. (2003). Early bioinformatics: The birth of a discipline—a personal view. *Bioinformatics* 19(17):2176–2190.

Promoter Element Discovery Tools

Aerts, S., Van Loo, P., Thijs, G., Moreau, Y., and De Moor, B. (2003). Computational detection of cis -regulatory modules. *Bioinformatics* 19(Suppl 2):II5–II14.

Brazma, A., Jonassen, I., Vilo, J., and Ukkonen, E. (1998). Predicting gene regulatory elements in silico on a genomic scale. *Genome Research* 8:1202–1215.

Bussemaker, H. J., Li, H., and Siggia, E. D. (2000). Building a dictionary for genomes: Identification of presumptive regulatory sites by statistical analysis. *Proc. Natl. Acad. Sci. USA* 97:10096–10100.

Birnbaum, K., Benfey, P. N., and Shasha, D. E. (2001). Cis element/transcription factor analysis (cis/TF): A method for discovering transcription factor/cis element relationships. *Genome Research* 11:1567–1573.

Chiang, D. Y., Brown, P. O., and Eisen, M. B. (2001). Visualizing associations between genome sequences and gene expression data using genome-mean expression profiles. *Bioinformatics* 17(Suppl 1):S49–S55.

Claverie, J.-M. (1999). Computational methods for the identification of differential and coordinated gene expression. *Hum. Mol. Genet.* 8:1821–1832.

Fujibuchi, W., Anderson, J. S. J., and Landsman, D. (2001). PROSPECT improves cis-acting regulatory element prediction by integrating expression profile data with consensus pattern searches. *Nucleic Acids Res.* 29:3988–3996.

Jensen, L. J., and Knudsen, S. (2000). Automatic discovery of regulatory patterns in promoter regions based on whole cell expression data and functional annotation. *Bioinformatics* 16:326–333.

Lawrence, C. E., Altschul, S. F., Boguski, M. S., Liu, J. S., Neuwald, A. F., and Wootton, J. C. (1993). Detecting subtle sequence signals: A Gibbs sampling strategy for multiple alignment. *Science* 262:208–214.

Liu, X., Brutlag, D. L., and Liu, J. S. (2001). BioProspector: Discovering conserved DNA motifs in upstream regulatory regions of co-expressed genes. *Pacific Symposium on Biocomputing* 6:127–138.[2]

Neuwald, A. F., Liu, J. S., and Lawrence, C. E. (1995). Gibbs motif sampling: Detection of bacterial outer membrane protein repeats. *Protein Science* 4:1618–1632.

Sheng, Q., Moreau, Y., and De Moor, B. (2003). Biclustering microarray data by Gibbs sampling. *Bioinformatics* 19(Suppl 2):II196–II205.

Spellman, P., Sherlock, G., Zhang, M., Lyer, V., Anders, K., Eisen, M., Brown, P., Botstein, D., and Futcher, B. (1998). Comprehensive identification of cell cycle-regulated genes of yeast *S. cerevisiae* by microarray hybridization. *Mol. Biol. Cell* 9:3273–3297.

Wolfsberg, T. G., Gabrielian, A. E., Campbell, M. J., Cho, R. J., Spouge, J. L., and Landsman, D. (1999). Candidate regulatory sequence elements for cell cycle-dependent transcription in *Saccharomyces cerevisiae*. *Genome Res.* 9:775–792.

Workman, C., and Stormo, G.D. (2000) ANN-Spec: A method for discovering transcription factor binding sites with improved specificity. *Pacific Symposium on Biocomputing 2000*.[3]

---

[2]Available online at http://psb.stanford.edu
[3]Available online at http://psb.stanford.edu

8

Automated Analysis, Integrated Analysis, and Systems Biology

Systems biology is basically the ability to study complex biological systems looking at all genes or all proteins, in terms of perturbations and model organisms. It's studying systems by looking at all the elements in the system rather than looking at things one at a time.

—Hood, 2002

If you have read the preceeding chapters you will have realized by now that there are quite many steps of an analysis and at each step you have a large choice of methods and parameters. The chapter on software will show free, public domain software to perform all the steps, but unless you are a computer scientist or bioinformatician the choices may seem bewildering to you. It is for that reason that we have ventured into the field of automated analysis. Is it possible to define a standard procedure for analysis, with a reasonable choice of methods, that will work on a majority of datasets? We believe so. While this is not necessarily the optimal way to analyse your data, it is a very fast and efficient first pass. Figure 8.1 shows the flow diagram of such a software system, GenePublisher (Knudsen, 2003). You will see that it looks much like the overview of analysis shown in Figure 2.1. It performs the analysis described in this book, but it does so completely automatically, without user intervention. It makes choices based on the data that you submit to it. Then it summarizes the results of all the analysis methods with the LaTeX report generation tool, and produces a PDF report in the form of a manuscript.

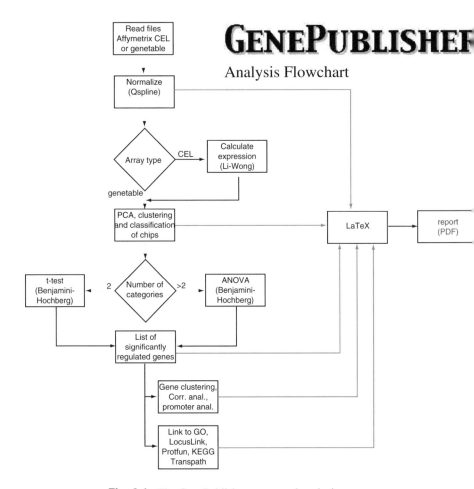

Fig. 8.1 The GenePublisher automated analysis system.

The output of GenePublisher is a good starting point for further analysis, for example verification of some of the predicted genes by another method such as RT-PCR. GenePublisher is available as a web server at www.cbs.dtu.dk/services/GenePublisher.

8.1   INTEGRATED ANALYSIS

GenePublisher also links the results to a number of biological databases. You want to know as much as possible about the significant genes: What is their exact biological function (Gene Ontology database)? Do they participate in some characterized signal transduction pathway (TRANSPATH database)? Are they enzymes in a characterized metabolic pathway (KEGG database)? Do they have any known transcription factor binding sites in their upstream regions that can account for the observed changes in expression (TRANSFAC database)? This is an *integrated analysis* where information from as many domains as possible are taken into account. You could also include information about known interactions between genes, either known protein-protein interactions from yeast two-hybrid experiments or known transcription factor binding to its known recognition site upstream of a gene (shown, e.g., via chromatin immunoprecipitation; Shannon, 2002).

This integrated analysis may help you in determining what goes on at a molecular level in your experiment. But it is rare that it results in one unifying hypothesis that can account for all the observed changes in gene expression. That is because most of our knowledge of molecular biology comes from the reductionistic approach that was common before the massively parallel methods such as arrays became available. Genes were studied one gene, one pathway, one protein at a time. It was assumed that everything else was held constant during the experiment. With the massively parallel approaches we see that this is not the case. Any change in expression of a gene results in changes in many other genes. And we rarely have the molecular knowledge to explain why all these changes occur – we lack knowledge of the *system*. For this, we need a *systems* approach to biology.

8.2   SYSTEMS BIOLOGY

Cells contain a large system of proteins and RNAs that interact with each other and control the transcription of genes. We know rather little of this system, but our knowledge is increasing rapidly as a results of the massive data gathering using parallel methods such as gene expression profiling, chromatin immunoprecipitation arrays, yeast two-hybrid assays, and so on. Many groups work at integrating these data into comprehensive models or maps of the underlying regulatory systems. As these models improve, they will improve our interpretation of microarray experiments. They will allow us to come up with models that explain more of the observed changes in expression than we are able to explain today.

It will also allow us to interpret microarray data in an entirely different way. In the statistical analysis we perform today to identify differentially regulated genes we assume that all genes are independent. That assumption is based on ignorance. We know they are dependent, but we do not know which genes

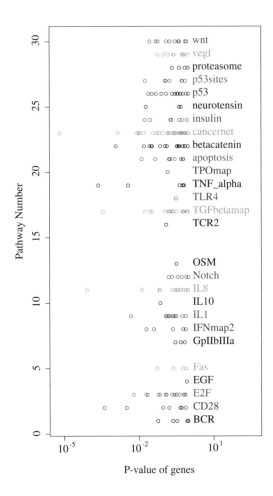

Fig. 8.2 Mapping of all genes on a human HuGeneFL chip to the signal tranduction pathways in the TRANSPATH database. For each pathway, the genes associated with it are plotted according to their *P*-value from the statistical analysis. That allows a view of whether more than one gene in a pathway is affected in the experiment you are studying.

depend on each other. Once we know more about the dependency of the genes we will use different statistical tests, not performed on individual genes in isolation but performed on networks of associated genes. It is evident from Figure 8.2 that merely looking at pathways captures very little, if any, of this association between genes. The genes in a pathway do not seem associated in their expression from this figure. We usually find significant differential expression for only a single gene in a pathway.

8.3   FURTHER READING

Grosu, P., Townsend, J. P., Hartl, D. L., and Cavalieri, D. (2002). Pathway Processor: A tool for integrating whole-genome expression results into metabolic networks. *Genome Research* 12(7):1121–1126.

Ideker, T., Thorsson, V., Ranish, J. A., Christmas, R., Buhler, J., Eng, J. K., Bumgarner, R., Goodlett, D. R., Aebersold, R., and Hood, L. (2001). Integrated genomic and proteomic analyses of a systematically perturbed metabolic network. *Science* 292:929–934.

Jenssen, T. K., Laegreid, A., Komorowski, J., and Hovig, E. (2001). A literature network of human genes for high-throughput analysis of gene expression. *Nature Genetics* 28:21–28.

Kanehisa, M., Goto, S., Kawashima, S., and Nakaya, A. (2002). The KEGG databases at GenomeNet. *Nucleic Acids Research* 30(1):42–46. [1]

Knudsen, S., Workman, C., Sicheritz-Ponten, T., and Friis, C. (2003). GenePublisher: Automated Analysis of DNA Microarray Data. *Nucleic Acids Research* 31(13):3471–3476. [2]

Krull, M., Voss, N., Choi, C., Pistor, S., Potapov, A., and Wingender, E. (2003). TRANSPATH: An integrated database on signal transduction and a tool for array analysis. *Nucleic Acids Research* 31(1):97–100. [3]

Masys, D. R., Welsh, J. B., Lynn Fink, J., Gribskov, M., Klacansky, I., and Corbeil, J. (2001). Use of keyword hierarchies to interpret gene expression patterns. *Bioinformatics.* 17:319–326.[4]

Matys, V., Fricke, E., Geffers, R., Gossling, E., Haubrock, M., Hehl, R., Hornischer, K., Karas, D., Kel, A. E., Kel-Margoulis, O. V., Kloos, D. U., Land, S., Lewicki-Potapov, B., Michael, H., Munch, R., Reuter, I., Rotert, S., Saxel, H., Scheer, M., Thiele, S., and Wingender, E. (2003). TRANSFAC: transcriptional regulation, from patterns to profiles. *Nucleic Acids Research* 31(1):374–378. [5]

Noordewier, M. O., and Warren, P. V. (2001). Gene expression microarrays and the integration of biological knowledge. *Trends. Biotechnol.* 19:412–415.

---

[1] Available at http://www.genome.ad.jp
[2] Available at http://www.cbs.dtu.dk/services/GenePublisher
[3] http://www.gene-regulation.com
[4] Web-based software available at http://array.ucsd.edu/hapi/
[5] http://www.gene-regulation.com

Rain, J. C., Selig, L., De Reuse, H., Battaglia, V., Reverdy, C., Simon, S., Lenzen, G., Petel, F., Wojcik, J., Schachter, V., Chemama, Y., Labigne, A., and Legrain, P. (2001). The protein-protein interaction map of *Helicobacter pylori. Nature* 409:211–215.

Shannon, M. F., and Rao S. (2002). Transcription. Of chips and ChIPs. *Science* 296(5568):666–669.

Tanabe, L., Scherf, U., Smith, L. H., Lee, J. K., Hunter, L., and Weinstein, J. N. (1999). MedMiner: An internet text-mining tool for biomedical information, with application to gene expression profiling. *BioTechniques* 27:1210–1217.[6]

Zhu, J., and Zhang, M. Q. (2000). Cluster, function and promoter: Analysis of yeast expression array. *Pacific Symposium on Biocomputing* 5:476–487.[7]

---

[6]Web version available at http://discover.nci.nih.gov/textmining/filters.html
[7]Available online at http://psb.stanford.edu

9

Reverse Engineering of Regulatory Networks

Cloning and studying individual genes is not enough. The sequence of the entire genome is not enough. Large-scale expression measurements and protein measurements are not enough. None of these sources of data will lead to a predictive understanding of the response of a living organisms to a stimulus unless they are integrated into a dynamical framework that can faithfully simulate the molecular interactions underlying life.

—Periwal, 2002

One gene can affect the expression of another gene by binding of the gene product of one gene to the promoter region of another gene. Looking at more than two genes, we refer to the *regulatory network* as the regulatory interactions between the genes.

If we have a large number of measurements of the expression level of a number of genes, we should be able to model or reverse engineer the regulatory network that controls their expression level. The problem can be attacked in two fundamentally different ways: using time-series data and using steady-state data of gene knockouts.

9.1 THE TIME-SERIES APPROACH

The expression level of a certain gene at a certain time point can be modeled as some function of the expression levels of all other genes at all previous time points. The problem is that you usually have many more genes than you have time points! That means that you have a dimensionality problem: there are too many parameters and too few equations to estimate them. If you have g genes, there are g^2 possible connections between them and you would need at least g^2 linearly independent equations to determine all of them.

Different solutions to this problem have been developed. One is the linear modeling approach by van Someren et al. (2000). They reduce the dimensionality of the problem by first removing all genes that do not show a significant change in expression through the experiment. Then they cluster the genes to group those that behave the same way. There is no reason mathematically to distinguish between two genes if you cannot distinguish their transcription response. Then they try to build a linear model of the remaining gene clusters. The basic linear model follows the assumption that the activity of a gene x equals the weighted sum of the activities of all N genes at the previous time point $(t - 1)$:

$$x_j(t) = \sum_{i=1}^{N} r_{i,j} x_i(t - 1),$$

where $r_{i,j}$ is a weight factor representing how gene i affects gene j, positively or negatively. Preliminary data indicate that they still need a few more experimental data points to solve the models exactly for the yeast cell cycle experiments than those that were available at the time.

Another solution to the dimensionality problem was developed by Holter et al. (2000), who used singular value decomposition (similar to principal component analysis, see Section 5.1) to reduce the dimensionality before solving the interaction matrix. That leaves fewer independent genes and makes it easier to find their interactions.

An entirely different approach is Bayesian networks, where the problem is simplified to one of genes that are up-regulated and genes that are down-regulated (Friedman et al., 2000). That still leaves a dimensionality problem, but they try to estimate probabilistic networks that fit the data and look for results that are common in different models that fit the same data. They have shown some success in extracting central regulatory pathways in yeast.

9.2 THE STEADY-STATE APPROACH

A particularly attractive approach is the steady-state model, where the effect of deleting a gene on the expression of other genes is measured. If the expression of gene b increases after deletion of gene a, it can be inferred that gene a repressed, either directly or indirectly, the expression of gene b. If the expression of gene b decreases after deletion of gene a, it can be inferred that gene a enhanced, either directly or indirectly, the expression of gene b. With a large DNA microarray, it is possible to determine all consequences of deletion of gene a.

Such results will give valuable information about the regulatory network in which the deleted gene is involved (Ideker et al., 2000, Kyoda et al., 2000). As compendiums of expression profiles of gene deletions become available (Hughes et al., 2000) the steady-state model is a very promising tool for extracting regulatory networks.

9.3 LIMITATIONS OF NETWORK MODELING

It must never be forgotten, however, that the genetic network approaches so far all ignore those regulatory interactions that take place at the protein-protein level. A lot of cellular regulation, for example of the cell cycle, takes place through phosphorylation and dephosphorylation of proteins. In the future, regulatory network models must include such information, for example, by inclusion of protein-protein interaction maps (Rain et al., 2001) determined by using the yeast two-hybrid assay. What is also needed is a way to combine prior biological knowledge of regulatory networks (Tanay and Shamir, 2001), information deduced from time-series experiments, and information deduced from steady-state experiments. If information from each can be represented as a matrix of interactions between genes, then the three matrices can be summed and the regulatory network deduced from that. The disadvantage of running the three methods independently, however, is that the solid information from prior knowledge and direct deletion in steady-state experiments can be useful in determining which time-series models best fit data from all three domains. Optimally, the information from prior knowledge and steady state models should be used when deriving time-series models.

The regulatory network can be visualized by drawing a box for each gene that has interactions above a cutoff threshold with other genes. Next all the interactions (above threshold) are drawn as lines between the boxes (line width can be scaled by interaction strength; positive and negative interactions can be distinguished by lines ending in an arrow and in a bar, respectively). For more than 100 interactions or so, this visualization quickly becomes unwieldy, and subnetworks have to be extracted and drawn.

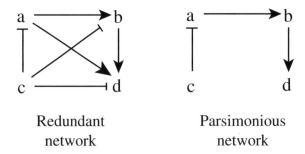

Redundant   Parsimonious
network   network

Fig. 9.1 Regulatory networks deduced from experimentally produced interaction matrix. Arrows mean positive regulation; bars mean negative regulation.

9.4 EXAMPLE 1: STEADY-STATE MODEL

Let us take an example of four genes, a, b, c, and d. When we delete gene a, we find that the expression of gene b and d decreases. We conclude that gene a has a stimulatory effect, directly or indirectly, on genes b and d. We can represent this information in an interaction matrix (Table 9.1). Note that there is a direction to each interaction. Rows represent genes that are deleted; columns represent genes whose expression is changed as a result. In the matrix we have included the results from two other deletion experiments: gene b, which led to a decrease in d, and gene c, which led to an increase in a, b, and d.

From this matrix we can now draw up a redundant genetic network that represents all the interactions between genes as arrows (positive regulation) or bars (negative regulation) (Figure 9.1).

This regulatory network is redundant in that it contains both direct and indirect regulations. There are several paths between two genes. What we now wish to deduce is the parsimonious network—the smallest and simplest network that is able to explain the experimental observations. If there is

Table 9.1 Interaction matrix between four genes.

Gene	Gene			
	a	b	c	d
a		$+$		$+$
b				$+$
c	$-$	$-$		$-$
d				

more than one path between two genes, we want to delete those that are not necessary to explain the results. That is achieved by eliminating for each pair of genes all but the longest path (involving most genes) if that path is still able to explain the regulatory effect observed.

Between genes c and b there are two possible paths that both have the same effect, so we remove the shortest path, the direct path between c and b. Between genes a and d there are two paths that both have the same effect, so we remove the shortest one, the direct path between a and d. Finally, between genes c and d, we remove the direct path between c and d. The resulting network is the simplest network that explains the data.

9.5 EXAMPLE 2: STEADY-STATE MODEL ON BACILLUS DATA

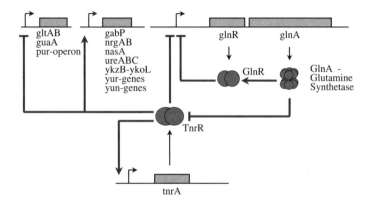

Fig. 9.2 Known regulatory network in *Bacillus subtilis*. Each line ending in a bar represents a negative regulatory effect. Each line ending in an arrow represents a positive regulatory effect. (Hanne Jarmer and Carsten Friis. See color plate.)

The approach described above was applied to knockout mutants in the regulation of nitrogen metabolism in *Bacillus subtilis* (Figure 9.2, Jarmer, et. al., (2002)). Expression data was filtered for significance through a *t*-test. Genes were clustered into groups that show the same response in all experiments. From the interaction matrix a redundant network was generated and reduced to a parsimonious network. Figure 9.3 shows the resulting network as output by a computer program. The perl program used is available upon request[1]

---

[1] See web companion site http://www.cbs.dtu.dk/steen/book.html

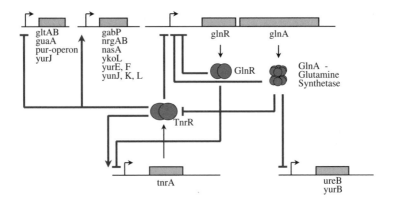

Fig. 9.3 Regulatory network reverse engineered from real steady-state data. Each line ending in a bar represents a deduced negative regulatory effect. Each line ending in an arrow represents a deduced positive regulatory effect. (Hanne Jarmer and Carsten Friis. See color plate.)

When compared to the known biological system shown in Figure 9.2, it is evident that the computer only missed the protein-protein interactions. But the computer discovered novel gene regulations in this system.

9.6 EXAMPLE 3: LINEAR TIME-SERIES MODEL

Let us try to deduce a network from time-series data as well. This is a little more complicated and involves some matrix algebra. Suppose we have conducted the experiment shown in Figure 9.4. At time zero we induce gene c with a substrate or other induction of its promoter. At times 1, 2, 3, and 4 we follow the expression level of gene c and three other genes, a, b, and d, and see how they change in response to the induction of gene c. We represent the expression of each gene at each time point as the logarithm (base 10) of the fold change relative to time zero. So gene a at time 0 is expressed at level $\log_{10}(1000/1000) = 0$, at time 1 is expressed at level $\log_{10}(1000/1000) = 0$, at time 2, 3, and 4 is expressed at level $\log_{10}(100/1000) = -1$. Gene c has expression levels 0, 1, 1, 1, 1 at time points 0, 1, 2, 3, and 4. We can put these logfold change numbers in an expression matrix (Table 9.2). To solve the regulatory network we plug into the formula mentioned above:

$$x_j(t) = \sum_{i=1}^{N} r_{i,j} x_i(t-1),$$

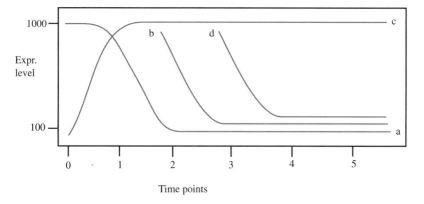

Fig. 9.4 Time-series experiment with four genes.

where $j = a, b, c, d$, and $i = a, b, c, d$, and $t = 1, 2, 3, 4$. This system of linear equations can be solved with standard techniques such as Gaussian elimination or singular value decomposition.

But we can illustrate the method by solving just the system of equations governing the regulation of gene a. At time $t = 4$, we have:

$$a(t = 4) = r_{b,a}b(t = 3) + r_{c,a}c(t = 3) + r_{d,a}d(t = 3).$$

Inserting the logfold values from our experiment time points 4 and 3, we get:

$$-1 = r_{b,a} - 1 + r_{c,a}1 + r_{d,a}0.$$

Likewise, we get for time points 3, 2, and 1:

Table 9.2 Expression matrix for four genes.

Gene	Time				
	0	1	2	3	4
a	0	0	-1	-1	-1
b	0	0	0	-1	-1
c	0	1	1	1	1
d	0	0	0	0	-1

Table 9.3   Directional interaction matrix between four genes.

	Gene			
Gene	a	b	c	d
a		$+$		
b				$+$
c	$-$			
d				

$$-1 = r_{b,a}0 + r_{c,a}1 + r_{d,a}0;$$

$$-1 = r_{b,a}0 + r_{c,a}1 + r_{d,a}0;$$

$$0 = r_{b,a}0 + r_{c,a}0 + r_{d,a}0.$$

Here we have four equations (two are identical) with three unknowns, and they can be solved with standard methods. It turns out that there is only one solution, $r_{c,a} = -1$ and $r_{b,a} = r_{d,a} = 0$. You can test this by inserting the solution into the four equations. So we have deduced from this set of equations that gene a is negatively regulated by gene c.

Likewise, we find that gene b is positively regulated by gene a, and gene d is positively regulated by gene b. We can summarize these findings in a directional interaction matrix (Table 9.3). The interactions are non-redundant, so it is easy to draw up the simplest network that satisfies the interaction matrix (Figure 9.5). This network is identical to the one deduced in Example 9.4.

It is rare, however, that the data are as well-behaved as in this hypothetical example. First, the time points must be sufficient to resolve unambiguously the order of events. So the separation between time points must be smaller than the time it takes to reach steady-state after induction of a gene. Second, the number of time points must be at least as large as the number of interactions between the genes studied.

Fig. 9.5   Regulatory network deduced from time-series interaction matrix. Arrows mean positive regulation; bars mean negative regulation.

9.7 FURTHER READING

Regulatory networks

Akutsu, T., Miyano, S., and Kuhara, S. (1999). Identification of genetic networks from a small number of gene expression patterns under the Boolean network model. *Pacific Symposium on Biocomputing* 4:17–28.[2]

Chen, T., He, H. L., and Church, G. M. (1999). Modeling gene expression with differential equations *Pacific Symposium on Biocomputing* 4:29–40.[3]

D'haeseleer, P., Wen, X., Fuhrman, S., and Somogyi, R. (1999). Linear modeling of mRNA expression levels during CNS development and injury *Pacific Symposium on Biocomputing* 4:41–52.[4]

Friedman, N., Linial, M., Nachman, I., and Pe'er, D. (2000). Using Bayesian networks to analyze expression data. *Proc. Fourth Annual International Conference on Computational Molecular Biology (RECOMB)* 2000.

Hartemink, A. J., Gifford, D. K., Jaakkola, T. S., and Young, R. A. (2001). Using graphical models and genomic expression data to statistically validate models of genetic regulatory networks. *Pacific Symposium on Biocomputing* 6:422–433.[5]

Hartemink, A. J., Gifford, D. K., Jaakkola, T. S., and Young R. A. (2002). Combining Location and Expression Data for Principled Discovery of Genetic Regulatory Network Models. *Pacific Symposium on Biocomputing* 2002:437–449.[6]

Holter, N. S., Maritan, A., Cieplak, M., Fedoroff, N. V., and Banavar, J. R. (2000). Dynamic modeling of gene expression data. *Proc. Natl. Acad. Sci. USA* 98:1693–1698.

Hughes, T. R., Marton, M. J., Jones, A. R., Roberts, C. J., and Stoughton, R., et. al. (2000). Functional discovery via a compendium of expression profiles. *Cell* 102:109–126.

---

[2] Available online at http://psb.stanford.edu
[3] Available online at http://psb.stanford.edu
[4] Available online at http://psb.stanford.edu
[5] Available online at http://psb.stanford.edu
[6] Available online at http://psb.stanford.edu

Ideker, T. E., Thorsson, V., and Karp, R. M. (2000). Discovery of regulatory interactions through perturbation: Inference and experimental design. *Pacific Symposium on Biocomputing* 5:305–316.[7]

Jarmer, H., Friis, C., Saxild, H. H., Berka, R., Brunak, S., and Knudsen, S. (2002). Inferring parsimonious regulatory networks in *B. subtilis*. *Pacific Symposium on Biocomputing* 2002. Poster presentation.

Kim, S., Dougherty, E. R., Bittner, M. L., Chen, Y., Sivakumar, K., Meltzer, P., and Trent, J. M. (2000) General nonlinear framework for the analysis of gene interaction via multivariate expression arrays. *J. Biomed. Opt.* 5:411–424.

Kim, S., Dougherty, E. R., Chen, Y., Sivakumar, K., Meltzer, P., Trent, J. M., and Bittner, M. (2000) Multivariate measurement of gene expression relationships. *Genomics* 15:201–209.

Kyoda, K. M., Morohashi, M., Onami, S., and Kitano, H. (2000). A gene network inference method from continuous-value gene expression data of wild-type and mutants. *Genome Informatics* 11:196–204.

Liang, S., Fuhrman S., and Somogyi, R. (1998). REVEAL, A general reverse engineering algorithm for inference of genetic network architectures. *Pacific Symposium on Biocomputing* 3:18–29.[8]

Maki, Y., Tominaga, D., Okamoto, M., Watanabe, S., and Eguchi, Y. (2001). Development of a system for the inference of large scale genetic networks. *Pacific Symposium on Biocomputing* 6:446–458.[9]

Pe'er, D., Regev, A., Elidan, G., and Friedman, N. (2001). Inferring subnetworks from perturbed expression profiles. *Bioinformatics* 17 (Suppl. 1):S215–S224.

Periwal, V., and Szallasi, Z. (2002). Trading "wet-work" for network. *Nature Biotechnology* 20:345–346.

Samsonova, M. G., and Serov, V. N. (1999). NetWork: An interactive interface to the tools for analysis of genetic network structure and dynamics. *Pacific Symposium on Biocomputing* 4:102–111.[10]

van Someren, E. P., Wessels, L. F. A., and Reinders, M. J. T. (2000). Linear modeling of genetic networks from experimental data. *Proc. ISMB* 2000:355–366.

[7]Available online at http://psb.stanford.edu
[8]Available online at http://psb.stanford.edu
[9]Available online at http://psb.stanford.edu
[10]Available online at http://psb.stanford.edu

Segal, E., Taskar, B., Gasch, A., Friedman, N., and Koller, D. (2001). Rich probabilistic models for gene expression. *Bioinformatics* 17(Suppl 1):S243–S252.

Shrager, J., Langley, P., and Pohorille, A. (2002). Guiding Revision of Regulatory Models with Expression Data *Pacific Symposium on Biocomputing* 2002:486–497.[11]

Szallasi, Z. (1999). Genetic network analysis in light of massively parallel biological data acquisition. *Pacific Symposium on Biocomputing* 4:5–16.[12]

Roberts, C. J., Nelson, B., Marton, M. J., Stoughton, R., Meyer, M. R., Bennett, H. A., He, Y. D. D., Dai, H. Y., Walker, W. L., Hughes, T. R., Tyers, M., Boone, C., and Friend, S. H. (2000). Signaling and circuitry of multiple MAPK pathways revealed by a matrix of global gene expression profiles. *Science* 287:873–880.

Tanay, A., and Shamir, R. (2001). Expansion on existing biological knowledge of the network: Computational expansion of genetic networks. *Bioinformatics* 17(Suppl 1):S270–S278.

Thieffry, D., and Thomas, R. (1998). Qualitative analysis of gene networks. *Pacific Symposium on Biocomputing* 3:77–88.[13]

Wagner, A. (2001). How to reconstruct a large genetic network from n gene perturbations in fewer than n(2) easy steps. *Bioinformatics* 17(12):1183–1197.

Wagner, A. (2002). Estimating coarse gene network structure from large-scale gene perturbation data. *Genome Research* 12(2):309–315.

Wahde, M., and Hertz, J. (2000). Coarse-grained reverse engineering of genetic regulatory networks. *Biosystems* 55:129–136.

Wahde, M., and Hertz, J. (2001). Modeling genetic regulatory dynamics in neural development. *J. Comput. Biol.* 8:429–442.

Weaver, D., Workman, C., and Stormo, G. (1999). Modeling regulatory networks with weight matrices. *Pacific Symposium on Biocomputing* 4:122–123.[14]

---

[11] Available online at http://psb.stanford.edu
[12] Available online at http://psb.stanford.edu
[13] Available online at http://psb.stanford.edu
[14] Available online at http://psb.stanford.edu

Wessels, L. F. A., Van Someren, E. P., and Reinders, M. J. T. (2001) A Comparison of genetic network models. *Pacific Symposium on Biocomputing* 6:508–519.[15]

Yeang, C. H., Ramaswamy, S., Tamayo, P., Mukherjee, S., Rifkin, R. M., Angelo, M., Reich, M., Lander, E., Mesirov, J., and Golub, T. (2001). Molecular classification of multiple tumor types. *Bioinformatics* 17(Suppl 1):S316–S322.

Yoo, C., Thorsson, V., and Cooper, G.F. (2002). Discovery of Causal Relationships in a Gene-Regulation Pathway from a Mixture of Experimental and Observational DNA Microarray Data. *Pacific Symposium on Biocomputing* 2002:498–509.[16]

---

[15] Available online at http://psb.stanford.edu
[16] Available online at http://psb.stanford.edu

10

Molecular Classifiers

Perhaps the most promising application of DNA microarrays for expression profiling is towards classification. In particular in medicine, where DNA microarrays may define profiles that characterise specific phenotypes (diagnosis), predict a patient's clinical outcome (prognosis), or predict which treatment is most likely to benefit the patient (tailored treatment).

The only limitation seems to be the fact that a sample of the diseased tissue is required for the chip. That limits the application to diseases that affect cells that can easily be obtained: blood disease where a blood sample can easily be obtained, or tumors where a biopsy is routinely obtained or the entire tumor is removed during surgery. Consequently, DNA microarrays have in the past few years been applied to almost any cancer type known to man, and in most cases it has been possible to distinguish clinical phenotypes based on the array alone. Where data on long-term outcome has been available, it has also been possible to predict that outcome to a certain extent using DNA arrays.

The key to the success of DNA microarrays in this field is that it is not necessary to understand the underlying molecular biology of the disease. Rather, it is a purely statistical exercise in linking a certain pattern of expression to a certain diagnosis or prognosis. This is called classification and it is a well established field in statistics from where we can draw upon a wealth of methods suitable for the purpose. This chapter will briefly explain some of the more common methods that have been applied to DNA microarray classification with good results.

10.1 FEATURE SELECTION

Feature selection is a very critical issue where it is easy to make mistakes. You can build your classifier using the expression of all genes on a chip as input or you can select the features (genes) that seem important for the classification problem at hand. As described below, it is easy to overfit the selected genes to the samples that you are basing the selection on. Therefore it is *crucial* to validate the classifier on a set of samples that were not used for feature selection.

There a many ways in which to select genes for your classifier – the simplest is to use the *t*-test or ANOVA described in this book to select genes differing significantly in expression between the different diagnosis categories or prognosis categories you wish to predict.

10.2 VALIDATION

If you have two cancer subtypes and you run one chip on each of them, can you then use the chip to classify the cancers into the two subtypes? With 6000 genes or more, easily. You can pick any gene that is expressed in one subtype and not in the other and use that to classify the two subtypes.

What if you have several cancer tissue specimens from one subtype and several specimens from the other subtype? The problem becomes only slightly more difficult. You now need to look for genes that all the specimens from one subtype have in common and are absent in all the specimens from the other subtype.

The problem with this method is that you have just selected genes to fit your data—you have not extracted a *general* method that will *classify any specimen of one of the subtypes that you are presented with after building your classifier.*

In order to build a *general* method, you have to observe several basic rules:

- Avoid overfitting data. Use fewer estimated parameters than the number of specimens that you are building your model on.

- Validate your method by testing it on an independent data set that was not used for building the model. (If your data set is very small, you can use *cross-validation* where you subdivide your data set into test and training several times. If you have ten examples, there are ten ways in which to split the data into a training set of nine and a test set of one. That is called a tenfold cross-validation. That is also called a leave-one-out cross validation or LOOCV.)

10.3 CLASSIFICATION SCHEMES

Most classification methods take as input points in space where each point corresponds to one patient or sample and each dimension in space corresponds to the expression of a single gene. The goal then becomes to classify a sample based on its position in space relative to the other samples and their known classes. As such, this method is related to the principal component analysis and clustering described elsewhere in this book. A key difference is that those methods are unsupervised, they do not use the information of the class relationship of each sample. Classification is supervised, the class relationship of each sample is used to build the classifier.

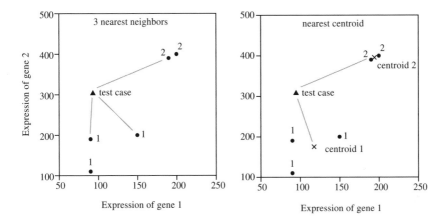

Fig. 10.1 Illustration of KNN (left) and nearest centroid classifier (right). The majority of the 3 nearest neighbors (left) belong to class 1, therefore we classify the test case as belonging to class 1. The nearest centroid (right) is that of class 1, therefore we classify the test case as belonging to class 1.

10.3.1 Nearest Neighbor

The simplest form of classifier is called a nearest neighbor classifier (Section 14.5.1.6) (Dudoit et al., 2000; Fix and Hodges, 1951). The general form uses k nearest neighbors and proceeds as follows: (1) plot each patient in space according to the expression of the genes; (2) for each patient, find the k nearest neighbors according to the distance metric you choose; (3) predict the class by majority vote, that is, the class that is most common among the k neighbors. If you use only odd values of k you avoid the situation of a vote tie. Otherwise, vote ties can be broken by a random generator. The value of k can be chosen by cross-validation to minimize the prediction error on a labeled test set.

If the classes are well separated in an initial principal component analysis (Section 5.4) or clustering, nearest neighbor classification will work well. If the classes are not separable by principal component analysis, it may be necessary to use more advanced classification methods, such as neural networks or support vector machines. The k nearest neighbor classifier will work both with and without feature selection. You can use either all genes on the chip or you can select informative genes with feature selection.

10.3.2 Nearest Centroid

Related to the nearest neighbor classifier is the nearest centroid classifier. Instead of looking at only the nearest neighbors is uses the centroids (center points) of all members of a certain class. The patient to be classified is assigned the class of the nearest centroid.

10.3.3 Neural Networks

If the number of examples is sufficiently high (between 50 and 100), it is possible to use a more advanced form of classification. Neural networks (Section 14.5.1.7) simulate some of the logic that lies beneath the way in which brain neurons communicate with each other to process information. Neural networks *learn* by adjusting the strengths of connections between them. In computer-simulated artificial neural networks, an algorithm is available for learning based on a learning set that is presented to the software. The neural network consists of an input layer where examples are presented, and an output layer where the answer, or classification category, is output. There can be one or more hidden layers between the input and output layer.

To keep the number of adjustable parameters in the neural network as small as possible, it is necessary to reduce the dimensionality of array data before presenting it to the network. Khan et al., (2001) used principal component analysis and presented only the most important principal components to the neural network input layer. They then used an ensemble of cross-validated neural networks to predict the cancer class of patients.

10.3.4 Support Vector Machine

Another type of classifier is the support vector machine (Brown et al., 2000; Dudoit et al., 2003), a machine learning approach. It is well suited to the dimensionality of array data. R code for implementing support vector machines can be found in the e1071 package at the R project web site (www.r-project.org).

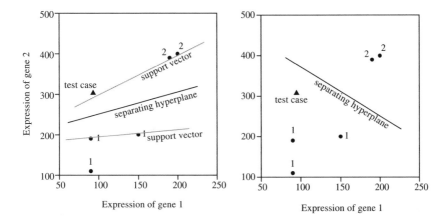

Fig. 10.2 Illustration of Support Vector Machine (left) and Neural Network classifier (right). Both define a separating hyperplane that can be defined in higher dimensions and can be more complex than what is shown here.

10.4 PERFORMANCE EVALUATION

There are a number of different measures for evaluating the performance of your classifier on an independent test set. First, if you have a binary classifier that results in only two classes (e.g., cancer or normal), you can use Matthews' correlation coefficient (Matthews, 1975) to measure its performance:

$$CC = \frac{(TP \times TN) - (FP \times FN)}{\sqrt{(TP + FN)(TP + FP)(TN + FP)(TN + FN)}},$$

where TP is the number of true positive predictions, FP is the number of false positive predictions, TN is the number of true negative predictions and FN is the number of false negative predictions. A correlation coefficient of 1 means perfect prediction, whereas a correlation coefficient of zero means no correlation at all (that could be obtained from a random prediction).

When the output of your classifier is continuous, such as that from a neural network, the numbers TP, FP, TN, and FN depend on the threshold applied to the classification. In that case you can map out the correlation coefficient as a function of the threshold in order to select the threshold that gives the highest correlation coefficient. A more common way to show how the threshold affects performance is, however, to produce a ROC curve (receiver operating characteristics). In a ROC curve you plot the sensitivity (TP/(TP+FN)) versus the false positive rate (FP/(FP+TN)). One way of comparing the performance

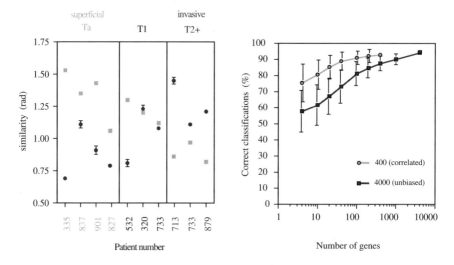

Fig. 10.3 Classifier of bladder cancers based on expression array. Left: Vector angle between patient and reference pool. Each three-digit number on the bottom refers to a patient. The angle between that patient and the two reference pools (squares, Ta pool; circles T2 pool) is indicated. The angle is always smallest (the similarity is greatest) to the pool with the same type of cancer. The intermediate type, T1, for which there is no reference pool, is sometimes more similar to one reference, sometimes more similar to another reference. Error bars have been added to show variation due to choice of reference pool, of which several were available. Right: the performance of the classifier as a function of the number of genes used for classification. Top curve: genes chosen among those 400 genes maximally covarying with the disease. Bottom curve: genes chosen at random from all 4000 genes detected as present in at least one patient. (Christopher Workman based on data from Thykjaer et al., (2001). See color plate.)

of two different classifiers is then to compare the area under the ROC curve. The larger the area, the better the classifier.

10.5 EXAMPLE I: CLASSIFICATION OF BLADDER CANCER SUBTYPES

As an example, our lab was faced with the problem of building a classifier that could categorize a bladder cancer as superficial or invasive based on a DNA chip test of a biopsy from the patient (Thykjaer et al., 2001). We only had biopsies from 10 patients. We decided to use a model without any estimated parameters at all. We simply measured the angle between the vector of all gene expression levels for each patient and the vectors of two reference samples of pools of superficial and invasive cancer. The angle was always smallest to the correct pool, because the vector angle distance was smallest

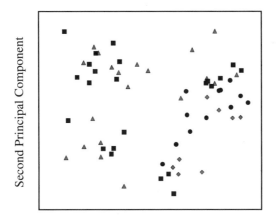

First Principal Component

Fig. 10.4   Principal component analysis of 63 small, round blue cell tumors. Different symbols are used for each of the four categories as determined by classical diagnostics tests. (See color plate.)

between samples from the same subtype (Figure 10.3). This approach was identical to the k nearest neighbor method with $k = 1$ and vector angle as distance measure. Since we had used no parameters to estimate this classifier, we expected it to be general. It was. When we received four new patient samples from our collaborators, they were correctly classified as well. The method was *validated*, although on a very small number of patients.

10.6   EXAMPLE II: CLASSIFICATION OF SRBCT CANCER SUBTYPES

Khan et al. (2001) have classified small, round blue cell tumors (SRBCT) into four classes using expression profiling and kindly made their data available on the World Wide Web[1]. We can test some of the classifiers mentioned in this chapter on their data.

First, we can try a k nearest neighbor classifier (see Section 14.5.1.6 for details). Using the full data set of 2000 genes, and defining the nearest neighbors in the space of 63 tumors by Euclidean distance, a $k = 3$ nearest neighbor classifier classifies 61 of the 63 tumors correctly in a leave-one-out cross-validation (each of the 63 tumors is classified in turn, using the remaining 62 tumors as a reference set).

---

[1] http://www.thep.lu.se/pub/Preprints/01/lu_tp_01_06_supp.html

We can also train a neural network (see Section 14.5.1.7 for details) to classify tumors into four classes based on principal components. Twenty feed-forward neural network are trained on 62 tumors, and used to predict the class of the 63rd tumor based on a committee vote among the twenty networks (this is a leave-one-out cross-validation). Figure 10.4 shows the first two principal components from a principal component analysis. The first ten principal components from a principal component analysis are used as the input for the neural network for each tumor, and four neurons are used for output, one for each category. Interestingly, two of the classes are best predicted with no hidden neurons, and the other two classes are best predicted with a hidden layer of two neurons. Using this setup, the neural networks classify 62 of the 63 tumors correctly. But of course, these numbers have to be validated on an independent, blind set as done by Khan et al. (2001).

10.7 SUMMARY

The most important points in building a classifier are these:

- Collect as many examples as possible and divide them into a training set and a test set.

- Use as simple a classification method as possible with as few adjustable (learnable) parameters as possible. Advanced methods (neural networks and support vector machines) require more examples for training than nearest-neighbor methods.

- Test the performance of your classifier on the independent test set. This independent test set must not have been used for selection of features (genes).

A more detailed mathematical description of the classification methods mentioned in this chapter can be found in Dudoit (2000).

10.8 FURTHER READING

Thykjaer, T., Workman, C., Kruhøffer, M., Demtröder, K., Wolf, H., Andersen, L. D., Frederiksen, C. M., Knudsen, S., and Ørntoft, T. F. (2001). Identification of gene expression patterns in superficial and invasive human bladder cancer. *Cancer Research* 61:2492–2499.

Class Discovery and Classification

Antal, P., Fannes, G., Timmerman, D., Moreau, Y., and De Moor, B. (2003). Bayesian applications of belief networks and multilayer perceptrons for ovarian tumor classification with rejection. *Artif Intell Med.* 29(1-2):39–60.

Bicciato, S., Pandin, M., Didone, G., and Di Bello, C. (2003). Pattern identification and classification in gene expression data using an autoassociative neural network model. *Biotechnol. Bioeng.* 81(5):594–606.

Brown, M. P. S., Grundy, W. N., Lin, D., Cristianini, N., Sugnet, C. W., Furey, T. S., Ares, M., and Haussler, D. (2000). Knowledge-based analysis of microarray gene expression data by using support vector machines *Proc. Natl. Acad. Sci. USA* 97:262–267.

Dudoit, S., Fridlyand, J., and Speed, T. P. (2000). Comparison of discrimination methods for the classification of tumors using gene expression data. Technical report #576, June 2000.[2]

Fix, E., and Hodges, J. (1951). Discriminatory analysis, nonparametric discrimination: Consistency properties. Technical report, Randolph Field, Texas: USAF School of Aviation Medicine.

Ghosh, D. (2002). Singular Value Decomposition Regression Models for Classification of Tumors from Microarray Experiments. *Pacific Symposium on Biocomputing* 2002:18–29.[3]

von Heydebreck, A., Huber, W., Poustka, A., and Vingron, M. (2001). Identifying splits with clear separation: A new class discovery method for gene expression data. *Bioinformatics* 17(Suppl 1):S107–S114.

Hastie, T., Tibshirani, R., Botstein, D., and Brown, P. (2001). Supervised harvesting of expression trees. *Genome Biol.* 2:RESEARCH0003.

Khan, J., Wei, J. S., Ringner, M., Saal, L. H., Ladanyi, M., Westermann, F., Berthold, F., Schwab, M., Antonescu, C. R., Peterson, C., and Meltzer, P. S. (2001). Classification and diagnostic prediction of cancers using gene expression profiling and artificial neural networks. *Nature Genetics* 7:673–679.

Matthews, B. W. (1975). Comparison of the predicted and observed secondary structure of T4 phage lysozyme. *Biochim. Biophys. Acta* 405:442–451.

Park, P.J., Pagano, M., and Bonetti, M. (2001). A nonparametric scoring algorithm for identifying informative genes from microarray Data. *Pacific Symposium on Biocomputing* 6:52–63.[4]

---

[2] Available at http://www.stat.berkeley.edu/tech-reports/index.html
[3] Available online at http://psb.stanford.edu
[4] Available online at http://psb.stanford.edu

Xiong, M., Jin, L., Li, W., and Boerwinkle, E. (2000). Computational methods for gene expression-based tumor classification. *Biotechniques* 29:1264–1268.

Yeang, C. H., Ramaswamy, S., Tamayo, P., Mukherjee, S., Rifkin, R. M., Angelo, M., Reich, M., Lander, E., Mesirov, J., and Golub, T. (2001). Molecular classification of multiple tumor types. *Bioinformatics* 17(Suppl 1):S316–S322.

11

The Design of Probes
for Arrays

11.1 SELECTION OF GENES FOR AN ARRAY

You may have so much knowledge of the molecular biology in a particular field that you already know the genes that you wish to include in a custom array. Say you are interested in a family of proteins, such as a particular class of receptors. If you are not sure that you know all the genes that are part of this family you can do a homology search or a Medline search. Both can be performed at the National Center for Biotechnology Information website[1]. The homology search is best performed starting with the amino acid sequence of one of the core family members and then using Psi-Blast (Altschul et al., 1997) to iteratively expand the family. The Medline search is done using PubMed by formulating keywords that are specific to your query and then seeing how well the resulting papers that are retrieved match those in which you are interested. By iterative reformulation of keywords you should be able to get a reasonable overview of the literature within a selected field— particularly when you look at the literature that has been cited by the relevant papers.

Another way of selecting genes for a spotted array is to use a commercial Affymetrix array to identify genes that are of interest to a particular problem. In that case the t-test or ANOVA (Section 4.6) are reliable tools to select relevant genes that differ significantly between conditions.

[1] http://www.ncbi.nlm.nih.gov

11.2 GENE FINDING

No matter what organism you are working with, there is a large fraction of genes that is not yet functionally characterized. They have been predicted either by the existence of a cDNA or EST clone with matching sequence, by a match to a homologous gene in another organism, or by *gene finding* in the genomic sequence. Gene finding uses computer software to predict the structure of genes based on DNA sequence alone (Guigo et al., 1992). Hopefully, they are marked as hypothetical genes by the annotator.

For certain purposes, for example, when designing a chip to measure all genes of a new microorganism, you may not be able to rely exclusively on functionally characterized genes and genes identified by homology. To get a better coverage of genes in the organism you may have to include those predicted by gene finding. Then it is important to judge the quality of the gene finding methods and approaches that have been used. While expression analysis may be considered a good method for experimental verification of predicted genes (if you find expression of the predicted gene it confirms the prediction), this method can become a costly verification if there are hundreds of false positive predictions that all have to be tested by synthesis of complementary oligonucleotides. A recent study showed that for *Escherichia coli* the predicted number of 4300 genes probably contains about 500 false positive predictions (Skovgaard et al., 2001). The most extreme case is the Archaea *Aeropyrum pernix*, where all open reading frames longer than 100 triplets were annotated as genes. Half of these predictions are probably false (Skovgaard et al., 2001).

Whether you are working with a prokaryote or a eukaryote, you can assess the quality of the gene finding by looking at which methods were used. If the only method used is looking for open reading frames as in the *A. pernix* case cited above, the worst prediction accuracy will result. Better performance is achieved when including codon usage (triplet) statistics or higher-order statistics (6th-order statistics, e.g., measure frequencies of hexamers). These frequencies are to some degree specific to the organism (Cole et al., 1998). Even better performance is obtained when including models for specific signals like splice sites (Brunak et al., 1990–1991), promoters (Knudsen, 1999; Scherf et al., 2000), and start codons (Guigo et al., 1992). Such signals are best combined within hidden Markov models which seem particularly well suited to the sequential nature of gene structure (Borodovski and McIninch, 1993; Krogh, 1997; Burge and Karlin, 1997).

11.3 SELECTION OF REGIONS WITHIN GENES

Once you have the list of genes you wish to spot on the array, the next question is one of cross-hybridization. How can you prevent spotting probes that are complementary to more than one gene? This question is of particular importance if you are working with a gene family with similarities in sequence. There is software available to help search for regions that have least similarity (determined by Blast; Altschul et al., 1990) to other genes. At our lab we have developed ProbeWiz[2] (Nielsen and Knudsen, 2002), which takes a list of gene identifiers and uses Blast to find regions in those genes that are the least homologous to other genes. It uses a database of the genome from the organism you are working with. Current databases available include: *Homo sapiens, Caenorhabditis elegans, Drosophila melanogaster, Arabidopsis thaliana, Saccharomyces cerevisae,* and *Escherichia coli.*

11.4   SELECTION OF PRIMERS FOR PCR

Once those unique regions have been identified, the probe needs to be designed from this region. It has been customary to design primers that can be used for polymerase chain reaction (PCR) amplification of a probe of desired length. ProbeWiz will suggest such primers if you tell it what length of the probe you prefer and whether you prefer to have the probe as close to the $3'$ end of your mRNA as possible. It will attempt to select primers whose melting temperature match as much as possible.

11.4.1   Example: Finding PCR Primers for Gene AF105374

GenBank accession number AF105374 (*Homo sapiens* heparan sulfate D-glucosaminyl 3-O-sulfotransferase-2) has been submitted to the web version of ProbeWiz, and the output generated if standard settings are used is given in Figure 11.1.

In addition to suggesting two primers for the PCR amplification, ProbeWiz gives detailed information on each of these primers and their properties, as well as a number of scoring results from the internal weighting process that went into selection of these two primers over others. The latter information may be useful only if you ask for more than one suggestion per gene and if you are comparing different suggestions.

11.5   SELECTION OF UNIQUE OLIGOMER PROBES

There is a trend in spotted arrays to improve the array production step by using long oligonucleotides (50 to 70 basepairs) instead of PCR products. It is also possible to use multiple 25 basepair probes for each gene as done by Affymetrix. Li and Stormo (2001) have run their DNA oligo (50–70 bases)

---

[2]Available in a web version at http://www.cbs.dtu.dk/services/DNAarray/probewiz.html

```
EST ID AF105374
Left primer sequence TGATGATAGATATTATAAGCGACGATG
Right primer sequence AAGTTGTTTTCAGAGACAGTGTTTTC
PCR product size 327
Primer pair penalty 0.6575 (Primer3)
```

	left primer	right primer
Position	1484	1810
Length	27	27
TM	59.8	60.4
GC %	33.3	33.3
Self annealing	6.00	5.00
End stability	8.60	7.30

Penalties:	Weighted	Unweighted
Homology	0	0
Primer quality	65.75	0.657
3'endness	158	158

Fig. 11.1   Output of ProbeWiz server upon submission of the human gene AF105374 (*Homo sapiens* heparan sulfate D-glucosaminyl 3-O-sulfotransferase-2).

prediction software on a number of complete genomes and made the resulting lists available online.[3]

We have developed a tool, OligoWiz[4], that will allow you to design a set of optimal probes (long or short oligos) for any organism for which you know the genome (Nielsen et al., 2003). In addition to selecting oligos that are unique to each gene, it also tries to assure that the melting temperature of the oligos selected for an array are as close to each other as possible (ΔT_m score) – if necessary by varying the length of the oligo.

In addition, OligoWiz assesses other properties of probes such as their position in the gene (distance to the $3'$ end, Position score) and the quality of the DNA sequence in the region where the probe is selected (GATC-only score). The user can assign a weight to the individual parameter scores associated with each probe that will affect the ranking of the probes that OligoWiz suggests for each gene.

---

[3] Available at http://ural.wustl.edu/~lif/probe.pl
[4] http://www.cbs.dtu.dk/services/OligoWiz

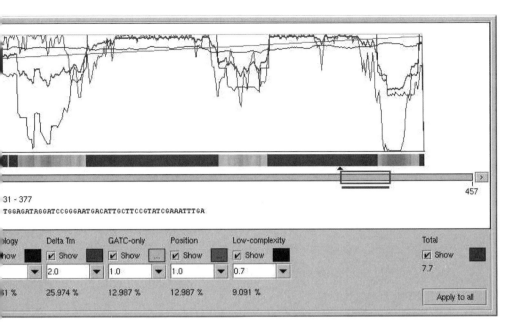

Fig. 11.2   Screenshot of part of the OligoWiz client program for designing oligonucleotide probes for a DNA microarray. The top window shows the distribution of the individual scores (defined in bottom window) over the entire gene.

11.6   REMAPPING OF PROBES

The assignment of probes to a gene is only as good as the bioinformatics used to design the probes. And the bioinformatics methods, as well as the annotation of the genomes based on which the probes are designed, improve all the time. For that reason a given probe-to-gene mapping can become outdated. For example, Affymetrix chips for human genes can be remapped using current human genome/transcriptome databases. We are in the process of submitting such a remapping of probes-to-genes to Bioconductor[5].

11.7   FURTHER READING

Primer and Oligo Probe Selection Tools

Li, F., and Stormo, G. D. (2001). Selection of optimal DNA oligos for gene expression arrays. *Bioinformatics* 17:1067–1076.

[5]http://www.bioconductor.org

Nielsen, H. B., and Knudsen, S. (2002). Avoiding cross hybridization by choosing nonredundant targets on cDNA arrays. *Bioinformatics* 18:321–322.

Nielsen, H. B., Wernersson, R., and Knudsen, S. (2003). Design of oligonucleotides for microarrays and perspectives for design of multi-transcriptome arrays. *Nucleic Acids Research* 31:3491–3496.

Varotto, C., Richly, E., Salamini, F., and Leister, D. (2001). GST-PRIME: A genome-wide primer design software for the generation of gene sequence tags. *Nucleic Acids Research* 29:4373–4377.

Rouillard, J. M., Herbert, C. J., and Zuker, M. (2002). OligoArray: genome-scale oligonucleotide design for microarrays. *Bioinformatics* 18(3):486–487.

Gene Finding

Altschul, S. F., Gish, W., Miller, W., Myers, E. W., and Lipman, D. J. (1990). Basic local alignment search tool. *J. Mol. Biol.* 215:403–410.[6]

Altschul, S. F., Madden, T. L., Schäffer, A. A., Zhang, J., Zhang, Z., Miller, W., and Lipman, D. J. (1997). Gapped BLAST and PSI-BLAST: A new generation of protein database search programs. *Nucleic Acids Res.* 25:3389–3402.[6]

Borodovsky, M., and McIninch, J. (1993). GeneMark: Parallel gene recognition for both DNA Strands. *Computers & Chemistry* 17:123–133.

Brunak, S., Engelbrecht, J., and Knudsen, S. (1990). Cleaning up gene databases. *Nature* 343:123.

Brunak, S., Engelbrecht, J., and Knudsen, S. (1990). Neural network detects errors in the assignment of mRNA splice sites. *Nucleic Acids Research* 18:4797–4801.

Brunak, S., Engelbrecht, J., and Knudsen, S. (1991). Prediction of human mRNA donor and acceptor sites from the DNA sequence. *Journal of Molecular Biology* 220:49–65.

Burge C., and Karlin, S. (1997). Prediction of complete gene structures in human genomic DNA. *Journal of Molecular Biology* 268:78–94.

Cole, S. T. et al. (1998). Deciphering the biology of *Mycobacterium tuberculosis* from the complete genome sequence. *Nature* 393:537–544.

---

[6]Available at http://www.ncbi.nlm.nih.gov/BLAST/

Guigo, R., Knudsen, S., Drake, N., and Smith. T. (1992). Prediction of gene structure. *Journal of Molecular Biology* 226:141–157.

Knudsen, S. (1999). Promoter2.0: For the recognition of PolII promoter sequences. *Bioinformatics* 15:356–361.[7]

Krogh, A. (1997). Two methods for improving performance of an HMM and their application for gene finding. *Proc. Fifth Int. Conf. on Intelligent Systems for Molecular Biology (ISMB)* Menlo Park, CA: AAAI Press, pp. 179–186.

Scherf, M., Klingenhoff, A., and Werner, T. (2000). Highly specific localization of promoter regions in large genomic sequences by PromoterInspector: A novel context analysis approach. *Journal of Molecular Biology* 297:599–606.

Skovgaard, M., Jensen, L. J., Brunak, S., Ussery, D., and Krogh, A. (2001). On the total number of genes and their length distribution in complete microbial genomes. *Trends Genet.* 17:425–428.

---

[7] Available as a web server at http://www.cbs.dtu.dk/services/Promoter/

12

Genotyping and Resequencing Chips

Up to this point this text has covered analysis of expression data. Chips for genotyping are also available. Instead of measuring mRNA, they measure DNA. Genotyping chips can detect mutations in genes. If you wish to screen each position in a gene for potential mutations, you are actually resequencing the gene. For example, a p53 chip is available from Affymetrix for detecting mutations in the DNA of human p53 tumor supressor gene. It does so with overlapping oligos that each are complementary to 20 base pairs of the *TP53* gene (Figure 12.1). Each oligo is present in 5 versions: the central nucleotide is either an A, C, G, T or is absent (a 1 bp deletion). Only one of these five oligos corresponds to the wild type (nonmutant) version of the *TP53* gene.

The PCR amplified, fragmented, fluorescently labeled DNA from a patient sample will hybridize to the complementary oligo for each position in the gene and it is then possible to read the sequence of the entire gene and determine whether it is equal to the wild type or not. There are still limitations of the accuracy of this determinination (Ahrendt et al., 1999; Wikman et al., 2000). Our lab has been working on neural network-based software that will improve the determination.

12.1 EXAMPLE: NEURAL NETWORKS FOR GENECHIP PREDICTION

Neural networks can be trained to predict DNA sequence based on the hybridization intensities measured on a chip designed for a specific gene (Spicker

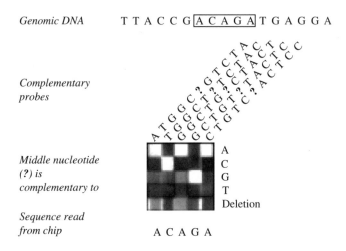

Fig. 12.1 p53 gene mutation analysis chip. Shown is a small area of the chip surface containing 25 different oligos designed to read the sequence of 5 nucleotides in the gene. The oligo probes (reduced to length 11 for clarity) for each position differ in their middle nucleotide (?) to account for the four possible bases as well the possibility of a one nucleotide deletion (right). By reading the intensity of the fluorescence at each oligo it is possible to read the DNA sequence. (Experiment by Jeppe Spicker.)

et al., 2002). It requires a large training set of DNA with accurately determined sequence.

Neural networks simulate some of the logic that lies beneath the way in which brain neurons communicate with each other to process information. Neural networks *learn* by adjusting the strengths of connections between them. In computer-simulated artificial neural networks, algorithms are available for learning based on a learning set that is presented to the software. In our case (Figure 12.2), the neural network consisted of an input layer which was presented with the measured fluorescence intensities for each of the ten probes used to detect each position: five with a central nucleotide of A, C, G, T, or deletion on the sense strand, and five with a central nucleotide of A, C, G, T, or deletion on the antisense strand. The output layer is then used to predict which nucleotide is present at the given position in the *TP53* gene. Between the input and output layers is a hidden layer which performs data processing. Such a neural network can then be trained by presenting examples matching input and output — *TP53* genes where the sequence has been determined by alternative means. When trained, it can predict the sequence based on a chip. This method is, however, sensitive to inhomogeneous samples where the sequence is ambiguous (Wikman, 2000), because it is a mixture of two alleles.

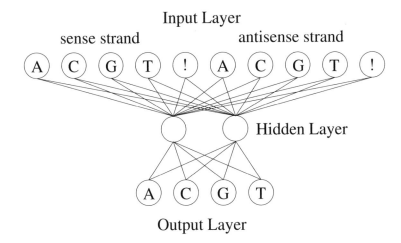

Fig. 12.2 Neural network for processing data from p53 chip shown in Figure 12.1. Shown are the units in the input layer, hidden layer, and output layer. The input for each position in the gene are the intensities of oligos whose central nucleotide is an A, C, G, T or deletion (!) for both the sense and the antisense strand. All units in one layer are connected with adjustable strengths to all units in the adjacent layer. (Figure by Jeppe Spicker.)

As an interesting aside, our neural network discovered an error in the labeling of input data during training. There were a few positions in the patient material that the neural network could not learn. Closer inspection revealed that it had detected known sites of polymorphisms that we had failed to label correctly in the training data (Spicker et al., 2002).

12.2 FURTHER READING

Ahrendt, S. A., Halachmi, S., Chow, J. T., Wu, L., Halachmi, N., Yang, S. C., Wehage, S., Jen, J., and Sidransky, D. (1999). Rapid p53 sequence analysis in primary lung cancer using an oligonucleotide probe array. *Proc. Natl. Acad. Sci. USA* 96:7382–7387.

Hacia, J. G. (1999). Resequencing and mutational analysis using oligonucleotide microarrays (Review). *Nat. Genet.* 21(1 Suppl):42–47.

Spicker, J. S., Wikman, F, Lu M. L., Cordon-Cardo C., Ørntoft, T. F., Brunak, S., and Knudsen, S. (2002). Neural network predicts sequence of TP53 gene based on DNA chip. *Bioinformatics* 18:1133–1134.

Wikman, F. P., Lu, M. L., Thykjaer, T., Olesen, S. H., Andersen, L. D., Cordon-Cardo, C., and Ørntoft, T. F. (2000). Evaluation of the performance of a p53 sequencing microarray chip using 140 previously sequenced bladder tumor samples. *Clin. Chem.* 46:1555–1561.

13

Experiment Design and Interpretation of Results

Design your experiment with the subsequent analysis and the inherent limitations of the technology in mind. The subsequent analysis usually consists of performing statistical tests on the genes on the array. If there are many genes on the the array you have the inherent problem of multiple testing which limits your ability to make inferences. Thus you need many replicates, and on a limited budget you get the most replicates for your money if you focus on comparing just two conditions. So try to identify the single comparison of conditions that answers most of what you want to know, and then replicate each of those conditions as many times as you can afford (three replicates at the very least). But because of the multiple testing problem, there will be both false positive and false negative conclusions from your experiment. You will need to verify the findings with an alternative experimental method such as RT-PCR.

13.1 FACTORIAL DESIGNS

It is also possible to investigate several factors simultaneously. In fact, this can be the only way to determine whether there is an interaction between factors. The two factors could be two different mutants. The 2 by 2 factorial design would then include an experiment with the wild type, an experiment with mutant 1, and experiment with mutant 2, and an experiment with the double mutant. You have to perform replicates of each of the four experiments, and an analysis of variance (ANOVA) then allows you to determine the effect of each of the factors. It also allows you to determine whether there is any interference between the two mutants. But you will need many more chips to investigate two factors than you need to investigate a single factor as described above. To aid in interpretation of the ANOVA results, it may be a good idea to use a parametrization designed for microarray experiments. See Diaz et al., (2001) for an example.

13.2 DESIGNS FOR TWO-CHANNEL ARRAYS

In the early days of spotted two-channel (Cy3 and Cy5) arrays investigators limited themselves to comparing channels only within arrays. That made experiment design complicated, because you needed a common reference to compare between arrays (Speed and Yang, 2002). We have taken a different approach. After observing that the fold changes resulting from comparisons between arrays were comparable to fold changes resulting from comparisons within arrays, and after observing that the two fold changes have similar standard deviations, we decided to treat the two channels as if they were separate arrays. That leaves no constraints on the design of the experiment beyond what is already mentioned above.

A condition for this approach is that the arrays are spotted in the same batch so the array to array variation is as small as possible. It is also important to use signal-dependent normalization before comparing between slides (arrays). Figure 13.1 shows a comparison of within-slide fold changes to between-slide fold changes for an experiment comparing a TnrA mutant in *Bacillus subtilis* to its wild type using spotted cDNA arrays with 12 replicates and 4111 genes (unpublished data). There are only a few genes out of the 4111 where there is a significant difference between the two ratios.

Another approach is to build an ANOVA model that takes into account all sources of variability and in the process treats the channels of two-channel arrays separately (Kerr et al., 2001; Wolfinger et al., 2001).

13.3 HYPOTHESIS DRIVEN EXPERIMENTS

Often DNA microarray experiments are performed without a hypothesis to be tested. We blindly search for genes that are differentially regulated and this blind search limits the usefulness of the experiment because of the multiple testing problem. If we instead had a hypothesis on one or more genes that we wish to test whether or not they are differentially expressed, the multiple testing problem suddenly disappears and we make substantial gains in power in the statistical tests. In fact, you would need much fewer replicates to test such a hypothesis, and you wouldn't need to verify the results with an independent method. But I guess the problem is that we know so little about molecular biology that we usually have no clue as to what to expect.

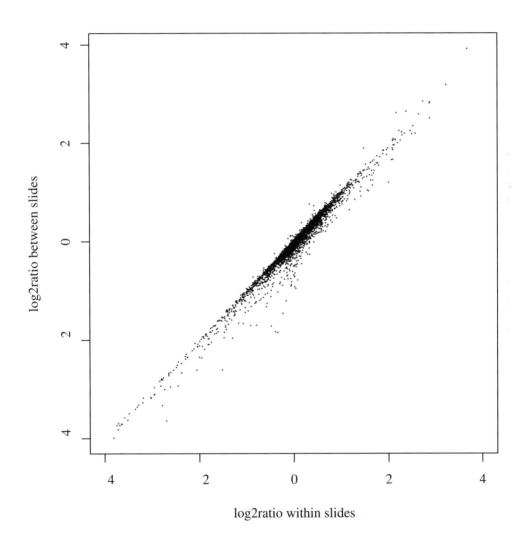

Fig. 13.1   Comparison of log ratios calculated within slides to log ratios calculated between slides. Both are averages of 12 replicate slides.

Hopefully this will change in the future as our knowledge of systems biology increases.

13.4   INDEPENDENT VERIFICATION

As we have seen in Section 4.6.3, there is a risk of making a false positive conclusion based on correct data. Furthermore, there are possibilities of making errors in the probe selection and probe design. For these reasons it usually required to verify important findings of differentially expressed genes with an independent method. Traditionally the choice of method has been quantitative PCR. This method is time-consuming, especially for a large number of genes. Fluorescent *in situ* hybridization (FISH) is even more attractive when dealing with heterogeneous tissue, because it shows the spatial and temporal expression pattern of the gene.

But if you need high-throughput verification of a number of genes, a high-throughput method is more attractive. If you have performed the first screening with Affymetrix GeneChips, you could use spotted arrays or *de novo* synthesized Febit Geniom arrays (Chapter 1.3) to verify a selected set of genes. If you design the probes *de novo*, and repeat the experiment from which your samples were extracted, the risk of making the same mistake twice is almost non-existent. Thus, you can consider the finding(s) independently verified.

13.5   INTERPRETATION OF RESULTS

A good place to start interpreting results is to look for possible outliers that may affect the entire analysis. Look at the MVA plots and their calculated variance. Are there any chips that deviate substantially from the rest? Is this confirmed by PCA and clustering of the chips? Is it confirmed by the lab notebook (low RNA yield, dubious electrophoresis results, poor amplification, image artifacts such as bubbles or scratches)? In that case try to remove that chip and see if the results of the statistical analysis improve.

If your chip includes an assessment of mRNA degradation by comparing the $5'$ and $3'$ ends of a gene, then that can also give you an indication of the quality of a particular sample.

After having removed outlying chips, if any, a good place to start the interpretation is to look at the most significant gene - the gene that shows differential regulation with the lowest *P*-value. This is the most certain conclusion from your experiment, so what is the function of this gene? Try to find out as much as possible from online databases such as LocusLink[1],

---

[1] http://www.ncbi.nlm.nih.gov/LocusLink/

GeneCards[2], PubMed[3], and so on. Then you continue to go down the list of genes ranked according to P-value. Is there anything in their function that relates to the condition that you are studying or relates to the function of other high ranking genes? Again, PubMed is a good place to answer this question, for example by searching for combinations of keywords.

How far should you go down the list of genes ranked according to significance? The estimated False Discovery Rate tells you how many false positives you expect after including the next gene on the list (by multiplying its P-value by the total number of genes on the chip). When the number of expected false positives becomes large compared to the number of genes you have accepted so far, you are venturing onto thin ice and any conclusions will be very weak. Proceed only if you are willing to test all genes with an alternative experimental method to sort out the true positives from the false positives.

Once you have settled for a list of genes there is one thing that is very important to remember. There will be false positives and false negatives on that list. You already have an estimate of the false positives, but calculating an estimate of the false negatives is more difficult. So the list of genes you see is far from the whole truth about your experiment. You see only a small window of the truth, and that window is spotted with false positives. It is for that reason you should be very careful about comparing your list of genes to lists of genes obtained in other experiments. Differences in the lists cannot be used to infer differences in the experiments, they can just as well reflect different samplings of false positives and false negatives. If you want to compare experiments, do so directly by comparing them in a t-test or other statistical test.

To get the whole truth you have to increase the power of the statistical test by increasing the number of replicates. As the number of replicates increases, you are asymptotically approaching the truth – the full list of differentially regulated genes without false positives. How many replicates it takes to come within a certain distance from the truth is difficult to say, for one thing it depends on your definition of differential regulation. But you certainly need more than 10 replicates.

13.6 LIMITATIONS OF EXPRESSION ANALYSIS

For all its strengths, it is important to keep in mind the limitations of expression analysis. First, expression analysis measures only the transcriptome. Important regulation takes place at the level of translation and enzyme activity. Those regulation effects are, at least for now, ignored in any analysis. In

[2]http://bioinfo.weizmann.ac.il/cards/
[3]http://www.ncbi.nlm.nih.gov

fact, significant signal transduction takes place at a protein to protein level. The only effect of such a signal transduction that you can observe in a gene expression experiment is any effect on gene expression that may be at the end of the signal transduction pathway.

Another issue that is largely ignored in current expression analysis is the effect of alternative splicing. To what extent are changes in observed signal from a particular messenger due to alternative splicing rather than due to a change in abundance? Current DNA microarrays for expression analysis have not been designed to distinguish between these two effects, largely because current knowledge of alternative splicing in the transcriptome is so limited.

In theory, the approach of using multiple probes per gene should be able to reveal alternative splicing if probes span an alternative splice junction. Thus, looking for changes in relative probe intensity within a gene might reveal alternative splicing (Hu et al., 2001) but it does not exclude the possibility of changes in cross-hybridisation for some of the probes within a gene.

Finally, keep in mind that mRNA is an unstable molecule. Messengers are programmed for enzymatic degradation and half-lives of messengers vary considerably. Messengers with very short half-lives may be difficult to extract in reproducible quantity. Thus, any regulation in expression of a gene with a very short half-life may be impossible to detect with statistical significance in a method that relies on reproducibility. Those unstable messengers will simply never pass a t-test.

In fact, unless the enzymatic degradation of messengers is stopped immediately after extraction of the sample (for example, by cooling in liquid nitrogen), there is not likely to be any unstable messengers left in the mRNA extraction. Those messengers will never be detected as present in your hybridization even though they may have been present inside the living cell.

For mRNA that has been extracted without proper care to immediately stop all degradation of messengers, all that is left is to analyze any changes in expression of stable messengers.

13.6.1 Relative Versus Absolute RNA Quantification

Most of this book has focused on relative changes in the abundance of a messenger RNA. Absolute quantification of copy number of mRNAs is a much harder task. You need to know how well each probe hybridizes to its target before you can use it to deduce anything about the absolute concentration of mRNA in the cell. One approach is to calibrate each probe set using known concentrations of their corresponding mRNA. That is labor intensive if you are working with many different mRNAs.

It is for that reason that most absolute analysis of mRNA has limited itself to the determination of whether a particular mRNA was present or not. Affymetrix in early version of their software made such a call based on empirically determined rules that take into account the number of PM/MM probe pairs which have a positive difference above a certain threshold, the number of PM/MM probe pairs which have a negative difference above a certain threshold, and the average log-ratio of all probe pairs, log(PM/MM). Schadt et al., (2000) have developed a statistical approach to the presence/absence determination.

Still, all you can say is whether a certain messenger RNA is above detection threshold or not. If it is significantly above, you may be able to say with high confidence that it is present. But if you do not detect a messenger, you cannot rule out that it is expressed below detection limit.

13.7 FURTHER READING

Diaz, E., Ge, Y., Yang, Y. H., Loh, K.C., Serafini, T. A., Okazaki, Y., Hayashizaki, Y., Speed, T. P., Ngai, J., and Scheiffele, P. (2002). Molecular analysis of gene expression in the developing pontocerebellar projection system. *Neuron* 36(3):417–34.

Kerr, M. K., Martin, M., and Churchill, G. A. (2000). Analysis of variance for gene expression microarray data. *Journal of Computational Biology* 7:819–37.

Speed, T. P., and Yang, Y. H. (2002) Direct versus indirect designs for cDNA microarray experiments Technical report #616, Department of Statistics, University of California, Berkeley.

Wolfinger, R. D., Gibson, G., Wolfinger, E. D., Bennett, L., Hamadeh, H., Bushel, P., Afshari, C., Paules, R. S. (2001). Assessing gene significance from cDNA microarray expression data via mixed models. *Journal of Computational Biology* 8(6):625–37.

Detection of Alternative Splicing with Affymetrix Chips

Hu, G. K., Madore, S. J., Moldover, B., Jatkoe, T., Balaban, D., Thomas, J., and Wang, Y. (2001). Predicting splice variant from DNA chip expression data. *Genome Research* 11:1237–1245.

Statistical Detection of Presence with Affymetrix Arrays

Schadt, E. E., Li, C., Su, C., and Wong, W. H. (2000). Analyzing high-density oligonucleotide gene expression array data. *J. Cell. BioChem.* 80:192–201.

14

Software Issues and Data Formats

In the future, sophisticated statistical, computational, and database methods may be as commonplace in Molecular Biology and Genetics as recombinant DNA is today.
—Pearson, 2001

Software for array data analysis is a difficult issue. There are commercial software solutions and there are noncommercial, public domain software solutions. In general, the commercial software lacks flexibility or sophistication and the non-commercial software lacks user-friendliness or stability.

Take Microsoft's Excel spreadsheet, for example. Many biologists use it for their array data processing, and indeed it can perform many of the statistical and numerical analysis methods described in this book. But there are pitfalls. First of all, commercial software packages can make assumptions about your data without asking you. As a scientist, you do not like to lose control over your calculations in that way. Second, large spreadsheets can become unwieldy and time-consuming to process and software stability can become an issue. Third, complicated operations require macro programming, and there are other, more flexible environments for programming.

You can perform the same types of analysis using non-commercial software. The best choice for microarray data analysis is the R statistics programming environment (www.r-project.org), where the Bioconductor consortium (www.bioconductor.org) has implemented most of the microarray analysis methods mentioned in this book. Numerous examples will be given later in this chapter. Extensive documentation on how to perform the analyses mentioned in this book are available from the Bioconductor web site.

There is, however, a steep learning curve. R is a very powerful language but to use it you must know about its objects and their structure. If this is not your cup of tea, there are other alternatives. You can use an automated, web-based approach such as GenePublisher[1], which is free and is based on Bioconductor. Or you can invest in a commercial software solution that integrates all analysis methods into one application.

14.1 STANDARDIZATION EFFORTS

There are several efforts underway to standardize file formats as well as the description of array data and underlying experiments. Such a standard would be useful for the construction of databases and for the exchange of data between databases and between software packages.

The Microarray Gene Expression Database Group[2] has proposed Minimum Information About a Microarray Experiment (MIAME) for use in databases and in publishing results from microarray experiments.

MGED is also behind the MAGE-ML standard for exchanging microarray data. MAGE-ML is based on the XML standard used on the World Wide Web. It uses tags, known from HTML, to describe array data and information.

14.2 DATABASES

A number of public repositories of microarray data have emerged. Among the larger ones are

- **ArrayExpress**
  European Bioinformatics Institute.
  http://www.ebi.ac.uk/arrayexpress/

- **Gene Expression Omnibus**
  National Institutes of Health.
  http://www.ncbi.nlm.nih.gov/geo/

- **GenePublisher**
  Technical University of Denmark.
  http://www.cbs.dtu.dk/services/GenePublisher/

[1] http://www.cbs.dtu.dk/services/GenePublisher
[2] www.mged.org

- **Stanford Microarray Database**
  Stanford University.
  http://genome-www5.stanford.edu/

14.3 STANDARD FILE FORMAT

The following standard file format is convenient for handling and analyzing expression data and will work with most software (including the web-based GenePublisher). Information about the experiment, the array technology, and the chip layout is not included in this format, and should be supplied in separate files. The file should be one line of tab-delimited fields for each gene (probe set):

Field 1: Gene ID or GenBank accession number

Field 2: (Optional) text describing function of gene

Fields 3 and up: Intensity values for each experiment including control, or logfold change for each experiment relative to the control

The file should be in text format, and not in proprietary, inaccessible formats that some commercial software houses are fond of using.

14.4 SOFTWARE FOR CLUSTERING

Once you have your data in a tab-delimited text format of Absolute expression values or Logfold values (see file format specification listed above), you can input it to the ClustArray software[3], which can do further data analysis for you:

- Rank genes based on vector length, covariance to disease progression, etc.

- Perform normalizations

- Cluster using a selection of different clustering algorithms and distance metrics

- Produce tree files to be visualized with drawtree or similar program

- Produce a distance matrix

- Visualize gene clusters with graphic Postscript file

---

[3] See web companion to this book http://www.cbs.dtu.dk/steen/book.html

Table 14.1 Expression readings of four genes in six patients.

Gene	N_1	N_2	A_1	A_2	B_1	B_2
	\multicolumn{6}{c}{Patient}
a	90	110	190	210	290	310
b	190	210	390	410	590	610
c	90	110	110	90	120	80
d	200	100	400	90	600	200

If ClustArray is installed on your system, type ClustArray -h to get a full list of all the options.

ClustArray is sensitive to the format of the input data, which must be tab separated and start with a line beginning with COLUMN_LABELS. ClustArray is available for Unix/Linux platforms.[3]

Other available software for clustering:

- Cluster. A widely used and user friendly software by Michael Eisen.[4] It is for Microsoft Windows only.

- GeneCluster. From the Whitehead/MIT Genome Center. Available for PC, Mac, and Unix.[5]

- Expression Profiler. A set of tools from the European Bioinformatics Institute that perform clustering, pattern discovery, and gene ontology browsing. Runs in a web server version and is also available for download in a Linux version.[6]

14.4.1 Example: Clustering with ClustArray

We will use a small standard example for clarity (Table 14.1). This can be converted into the format required by ClustArray with the following awk script:

```
awk 'BEGIN{printf"COLUMN_LABELS\t"} \
{for(i = 1; i < NF; i + +) printf"%s\t",$i; printf"%s\n",$NF}' \
example >exampleCE
```

Now we can use ClustArray to hierarchically cluster genes based on vector angle distance:

```
ClustArray -f exampleCE -m exampleCE.mtx \
-t exampleCE.new -Ci 1 -Ca 0 -Cd 2   -Cm 2 -Cc 1
```

---

[4]http://rana.lbl.gov
[5]http://www-genome.wi.mit.edu/MPR/
[6]http://ep.ebi.ac.uk

Table 14.2 Expression readings of four genes in six patients.

Gene	Patient					
	N_1	N_2	A_1	A_2	B_1	B_2
a	90	110	190	210	290	310
b	190	210	390	410	590	610
c	90	110	110	90	120	80
d	200	100	400	90	600	200

The software has produced two files: exampleCE.mtx contains the distance matrix and exampleCE.new contains a Newick description of the resulting tree. Most tree viewers (Phylip's drawgram, TreeView, etc.) can read this tree format and draw the tree for you. The resulting drawing is shown in Figure 6.6.

14.5 SOFTWARE FOR STATISTICAL ANALYSIS

Microsoft Excel has several statistics functions built in, but an even better choice is the free, public-domain R package which can be downloaded for Unix/Linux, Mac, and Windows systems.[7] This software can be used for t-test, ANOVA, principal component analysis, clustering, classification, neural networks, and much more. The Bioconductor[8] consortium has implemented most of the array specific normalization and statistics methods described in this book.

14.5.1 Example: Statistical Analysis with R

In this section we will use a small standard example for clarity (Table 14.2). There are four genes, each measured in six patients, which fall into three categories: normal (N), disease stage A, and disease stage B. That means that each category has been *replicated* once. The data follow the standard format except that there is no function description.

Next you boot up R (by typing the letter R at the prompt) and read in the file:

```
dataf  <- read.table("example")
```

In the web companion to this book you can find the example and code for copy-paste to your own computer[9]

---

[7]http://www.r-project.org
[8]http://www.bioconductor.org
[9]http://www.cbs.dtu.dk/steen/book.html

14.5.1.1 The t-test Here is how you would perform a *t*-test to see if genes differ significantly between patient category A and patient category B:

```
# load library for t-test:
library(ctest)
# t-test function:
get.pval.ttest <- function(dataf,index1,index2,
   datafilter=as.numeric){
   f <- function(i) {
      return(t.test(datafilter(dataf[i,index1]),
         datafilter(dataf[i,index2])))$p.value)
   }
   return(sapply(1:length(dataf[,1]),f))
}
# call function with our data:
pVal.ttest <- get.pval.ttest(dataf,3:4,5:6)
# print results on screen (only for a small dataset like this)
print(cbind(dataf,pVal.ttest))
        V1  V2  V3  V4  V5  V6  V7 pVal.ttest
1 gene_a  90 110 190 210 290 310 0.01941932
2 gene_b 190 210 390 410 590 610 0.00496281
3 gene_c  90 110 110  90 120  80 1.00000000
4 gene_d 200 100 400  90 600 200 0.60590011
# sort on P-value and write to file:
orders <- order(pVal.ttest)
ordered.data <- cbind(dataf[orders,],pVal.ttest[orders])
write(t(as.matrix(ordered.data)),
   ncolumns=length(dataf)+1,
   file = "ttest.out")
q(save="no")
```

The default call to the t.test function assumes unequal variance between the two populations (Welch's *t*-test).

14.5.1.2 Wilcoxon You can perform the Wilcoxon test instead of the *t*-test by replacing the call to the t.test function above with a call to the wilcox.test function.

14.5.1.3 ANOVA Here is how you would perform an ANOVA on the example to test for genes that differ significantly in at least one of the three categories N, A, and B:

```
# Specify categories and columns holding AvgDiff data:
Categories <- as.factor(c("0","0","A","A","B","B"))
indexAvgDiff <- 1:6
# ANOVA function:
get.pval.anova <-
   function(dataf,indexAll,Categories,
   datafilter=as.numeric){
   Categories <- as.factor(Categories)
   f <- function(i) {
      return(summary(
         aov(datafilter(dataf[i,indexAll]) ~ Categories)
      )
      [[1]][4:5][[2]][1])
```

```
    }
    return(sapply(1:length(dataf[,1]),f))
  }
# call the function with our data:
  pVal.anova <- get.pval.anova(dataf,indexAvgDiff,Categories)
# print results on screen (only for a small dataset like this)
  print(cbind(dataf,pVal.anova))
         V1  V2  V3  V4  V5  V6  V7   pVal.anova
  1 gene_a  90 110 190 210 290 310 0.0017965439
  2 gene_b 190 210 390 410 590 610 0.0002283540
  3 gene_c  90 110 110  90 120  80 1.0000000000
  4 gene_d 200 100 400  90 600 200 0.5560965577
# sort on P-value and write results to file:
  orders <- order(pVal.anova)
  ordered.data <- cbind(dataf[orders,],pVal.anova[orders])
  write(t(as.matrix(ordered.data)),
   ncolumns=length(dataf)+1,
   file = "ANOVA.out")
  q(save="no")
```

In general, it is best to perform statistical tests on the raw expression values instead of fold change. If you are working with spotted arrays where there is much variation between slides it may be better to perform the statistical test on the fold change derived from the red and green channels of each slide. Try to perform the statistical test on both absolute values and fold change and compare the results. Baldi and Long (2001) advocate log-transformation of data before statistical analysis.

14.5.1.4 PCA You can perform a principal component analysis of the same data as follows (note that the first row of the data must contain labels of all patients, but no label for the gene identifier):

```
library(mva)
dataf  <- read.table("example")
pca <- princomp(dataf)
summary(pca)
plot(pca)
biplot(pca)
```

This will produce the plot shown in Figure 5.2 in Section 5.1. The plot shown in Figure 5.3 was produced by transposing the data matrix with the command t(dataf), but PCA may not work for large transposed matrices.

14.5.1.5 Correspondence Analysis You can perform a correspondence analysis much the same way as you perform a principal component analysis:

```
library(MASS)
library(mva)
dataf  <- read.table("example",header=T)
cs <- corresp(dataf,nf=2)
plot(cs)
```

14.5.1.6 Classification You can perform k nearest neighbor classification with cross-validation using builtin functions in the standard distribution of R. This is the calculation that was performed for the classifier in Section 10.6:

```
library(class)
library(mva)
dataf  <- read.table("datafile",header=T)
tposed <- t(dataf)
knn.targets <- factor( c(rep("E", 23), rep("B", 8),
                 rep("N", 12), rep("R", 20)) )
knn.cv(tposed, knn.targets, k=3, prob=TRUE)
E E E E E E E E E E E E E E E E E E E E E E E
B B B B B B B B N N N N N N N N N N N N R R R N
R R R R R E R R R R R R R R R
```

The predicted classes (last two lines) differ from those assigned (knn.targets) for two tumors, so this classification is 97% correct.

14.5.1.7 Neural Networks The R package has a function for training a feed-forward neural network and using the trained network to predict the class of unlabeled samples. These calculations were performed for the neural network results shown in Section 10.6 with leave-one-out cross-validation:

```
library(class)
library(nnet)
pca <- princomp(tposed,cor=TRUE,scores=TRUE)
canc <- pca$scores[,1:10]
targets <- class.ind( c(rep("E", 23), rep("B", 8),
                rep("N", 12), rep("R", 20)) )
f <- function(samp) {
    b <- c(0,0,0,0)
    for(i in 1:20) {
        trainednet <- nnet(canc[-samp,], targets[-samp,],
             size=2,skip=FALSE,trace=FALSE, maxit=300)
        a <- max.col(predict.nnet(trainednet, canc[samp,]))
        b[a] <- b[a] + 1
    }
    print(b)
}
lapply(1:63,f)
```

14.5.2 The affy Package of Bioconductor

We have contributed to the affy package (Gautier et al., 2003) at Bioconductor which, in addition to the above-mentioned statistical analyses, can replace Affymetrix GeneChip software by reading CEL files directly and calculating expression values using the Li-Wong method as well as other methods. It can normalize the data using a range of signal-dependent normalization methods. Here is an example of how the functions in the affy package can be called to read and analyse a batch of Affymetrix CEL files (start by loading R version 1.7.1 or higher):

```
# use version 1.2.30 or higher of the affy package:
  library(affy)
# read CEL files:
  data <- read.affybatch("day_7a_amp.CEL", "day_7b_amp.CEL",
            "day_7c_amp.CEL", "day_7a_HIV_amp.CEL",
            "day_7b_HIV_amp.CEL", "day_7c_HIV_amp.CEL")
  # background correct, normalize, and condense:
  affy.es <- expresso(data, bgcorrect.method="rma2",
              normalize.method = "qspline",
              pmcorrect.method="pmonly",
              summary.method ="liwong")
  Matrix <- exprs(affy.es)
# t-test function:
  get.pval.ttest <- function(dataf,index1,index2,
            datafilter=as.numeric)
      f <- function(i)
          return(t.test(datafilter(dataf[i,index1]),
            datafilter(dataf[i,index2])))$p.value)

      return(sapply(1:length(dataf[,1]),f))

# perform t-test:
  pValues <- get.pval.ttest(Matrix,1:3,4:6)
# write ranked results to file:
  orders <- order(pValues)
  ordered.data <- cbind(rownames(Matrix)[orders],Matrix[orders,],
                pValues[orders])
  write(t(as.matrix(ordered.data)), ncolumns=1+ncol(Matrix)+1,
        file = "Pvalues.abs")
```

The input data files are available in the web companion for this book[10]. The affy package can, counterintuitively, be used to analyse data from spotted arrays and other platforms as well. Here is an example using an input file in the standard tab-delimited text file format described above with no annotation field and no header (start by loading R version 1.7.1 or higher):

```
  library(affy)
# read file:
  dataf <- read.table("input.file", row.names=1, sep="""^,
            header=F, comment.char = "")
# normalize using qspline:
  normdata <- normalize.qspline(dataf, na.rm=TRUE)
  # t-test function:
  get.pval.ttest <- function(dataf,index1,index2,
            datafilter=as.numeric)
      f <- function(i)
          return(t.test(datafilter(dataf[i,index1]),
            datafilter(dataf[i,index2])))$p.value)

      return(sapply(1:length(dataf[,1]),f))

# perform t-test:
  pValues <- get.pval.ttest(normdata,1:12,13:24)
```

---

[10]http://www.cbs.dtu.dk/steen/book.html

```
# write ranked results to file:
orders <- order(pValues)
ordered.data <- cbind(rownames(dataf)[orders],normdata[orders,]
                pValues[orders])
write(t(as.matrix(ordered.data)), ncolumns=1+ncol(normdata)+1,
    file = "Pvalues.abs")
```

14.5.3 Commercial Statistics Packages

The commercial statistics packages SAS, SPSS, and S-PLUS include the statistical functions described in this section as well.

14.6 SUMMARY

The R programming language with the Bioconductor packages offer the best solution for data analysis. They run on any PC, Mac or Unix system. With such a setup your possibilities for data mining and discovery are limited only by your imagination (and, perhaps, by your programming skills).

14.7 FURTHER READING

Baldi, P., and Long, A. D. (2001). A Bayesian framework for the analysis of microarray expression data: Regularized *t*-test and statistical inferences of gene changes. *Bioinformatics* 17:509–519.[11]

Gautier, L., Cope, L., Bolstad B.M., and Irizarry R. A. (2003) affy - Analysis of Affymetrix GeneChip data at the probe level. Bioinformatics, in press.

Parmigiani, G., Garrett, E. S., Irizarry, R. A., and Zeger, S. (Editors), (2003) *The Analysis of Gene Expression Data* Berlin: Springer Verlag.

Pearson, W. R. (2001). Training for Bioinformatics and Computational Biology (Editorial). *Bioinformatics* 17:761–762.

R

Online manuals available at http://www.r-project.org.

Venables, W. N. and Ripley, B. D. (1999). *Modern Applied Statistics with S-PLUS*, 3rd ed. New York: Springer.

---

[11]Accompanying web page at http://visitor.ics.uci.edu/genex/cybert/

Appendix A
Web resources:
Commercial Software
Packages

The previous chapter introduced the R and Bioconductor environments that can perform all the analyses described in this book. Such an environment is most suited for bioinformatics work because it has unlimited flexibility. No DNA array analysis tasks are identical and you need to tailor each analysis to the problem at hand.

However, not every biologist wants to deal with R programming. There are alternatives. Complete software packages for performing analysis of DNA microarray data are available for Microsoft Windows platforms and other platforms. I will briefly mention some of them and describe the functionality that they currently include.

- **Affymetrix Data Mining Tool**
  This software offers statistical analysis as well as clustering and visualization of Affymetrix GeneChip data. It is integrated with a laboratory information management system to offer data management of large volumes of chip data. The price is targeted at large pharmaceutical

companies.
http://www.affymetrix.com

- **Affymetrix NetAffx**
  An online resource freely accessible to Affymetrix customers, it links probes on Affymetrix GeneChips to public and proprietary databases, allowing users to integrate data analysis from their expression experiments.
  http://www.netaffx.com

- **Biomax Gene Expression Analysis Suite**
  Clusters genes and links genes in selected clusters to metabolic pathways. Protein interaction networks of co-expressed genes are identified from a database of protein interactions for the respective organism.
  http://www.biomax.de

- **GeneData Expressionist**
  A software system for organizationwide analysis of expression data from a variety of platforms including Affymetrix and spotted cDNA arrays and filters. Expressionist performs background subtraction, scaling, statistical tests, clustering, data visualization, and promoter searching. The software is targeted at biotech industries.
  http://www.genedata.com/

- **GeneSifter.Net**
  A web-based software system for management and analysis of microarray data.
  http://www.genesifter.net/

- **Informax Xpression NTI**
  Imports expression data in a variety of formats, performs normalization, clustering, and graphical representation of data. Informax also offer ArrayPro Analyzer to process spotted array images to expression values.
  http://www.informaxinc.com

- **Invitrogen Corporation ResGen Pathways**
  A comprehensive set of tools for the analysis of differential gene expression using microarray data. Image analysis, statistical analysis, clustering, pathway analysis and linking to databases.
  http://pathways.resgen.com/

- **Lion Bioscience arraySCOUT**
  Enterprisewide expression data analysis designed to handle large volumes of expression data. The software includes statistical analysis, clustering, and visualization tools. Includes database of gene annotation.
  http://www.lion-bioscience.com/

- **GeneLinker**
  GeneLinker is shrink-wrap software from Predictive Patterns Software Inc. It performs standard microarray data analysis.
  http://www.predictivepatterns.com/

- **Rosetta Resolver Gene Expression Data Analysis System**
  The Rosetta Resolver is an enterprisewide gene expression data analysis system that combines analysis software, a high capacity database, and high-performance server hardware to enable users to store, retrieve, and analyze large volumes of gene expression data. It is accessed from PC clients. The software includes statistical analysis, five different clustering algorithms, and numerous visualization tools. The price is targeted at very large pharmaceutical companies.
  http://www.rosettabio.com//

- **Silicon Genetics GeneSpring**
  Performs clustering, multiple visualizations, and annotation of expression data from multiple sources. It includes a regulatory sequence search algorithm. The price is targeted at smaller research groups.
  http://www.sigenetics.com

- **Spotfire**
  A multipurpose tool for analyzing and visualizing data.
  http://www.spotfire.com/

Most of these software packages use Microsoft Windows. GeneSpring is available in a Macintosh version as well.

References

1. Aerts, S., Van Loo, P., Thijs, G., Moreau, Y., and De Moor, B. (2003). Computational detection of cis -regulatory modules. *Bioinformatics* 19(Suppl 2):II5–II14.

2. Affymetrix (1999). *GeneChip Analysis Suite*. User Guide, version 3.3.

3. Affymetrix (2000). *GeneChip Expression Analysis*. Technical Manual.

4. Ahrendt, S. A., Halachmi, S., Chow, J. T., Wu, L., Halachmi, N., Yang, S. C., Wehage, S., Jen, J., and Sidransky, D. (1999). Rapid p53 sequence analysis in primary lung cancer using an oligonucleotide probe array. *Proc. Natl. Acad. Sci. USA* 96:7382–7387.

5. Akutsu, T., Miyano, S., and Kuhara, S. (1999). Identification of genetic networks from a small number of gene expression patterns under the boolean network model. *Pacific Symposium on Biocomputing* 4:17–28.[1]

6. Albert, T. J., Norton, J., Ott, M., Richmond, T., Nuwaysir, K., Nuwaysir, E. F., Stengele, K. P., and Green, R. D. (2003). Light-directed $5' \rightarrow 3'$ synthesis of complex oligonucleotide microarrays. *Nucleic Acids Research* 31(7):e35.

---

[1] Available online at http://psb.stanford.edu

7. Alter, O., Brown, P. O., and Botstein, D. (2000). Singular value decomposition for genome-wide expression data processing and modeling. *Proc. Natl. Acad. Sci. USA* 97:10101–10106.

8. Altschul, S. F., Gish, W., Miller, W., Myers, E. W., and Lipman, D. J. (1990). Basic local alignment search tool. *J. Mol. Biol.* 215:403–410.[2]

9. Altschul, S. F., Madden, T. L., Schäffer, A. A., Zhang, J., Zhang, Z., Miller, W., and Lipman, D. J. (1997). Gapped BLAST and PSI-BLAST: A new generation of protein database search programs. *Nucleic Acids Res.* 25:3389–3402.[2]

10. Antal, P., Fannes, G., Timmerman, D., Moreau, Y., and De Moor, B. (2003). Bayesian applications of belief networks and multilayer perceptrons for ovarian tumor classification with rejection. *Artif Intell Med.* 29(1-2):39–60.

11. Audic, S., and Claverie, J. M. (1997). The significance of digital gene expression profiles. *Genome Res.* 7:986–995.

12. Baggerly, K. A., Coombes, K. R., Hess, K. R., Stivers, D. N, Abruzzo, L. V., and Zhang, W. (2001). Identifying differentially expressed genes in cDNA microarray experiments. *Journal of Computational Biology* 8(6):639–659.

13. Baldi, P., and Long, A. D. (2001). A Bayesian framework for the analysis of microarray expression data: Regularized t-test and statistical inferences of gene changes. *Bioinformatics* 17:509–519.[3]

14. Baugh, L. R., Hill, A. A., Brown, E. L., and Hunter, C. P. (2001). Quantitative analysis of mRNA amplification by in vitro transcription. *Nucleic Acids Research* 29:E29.

15. Baum, M., Bielau, S., Rittner, N., Schmid, K., Eggelbusch, K., Dahms, M., Schlauersbach, A., Tahedl, H., Beier, M., Guimil, R., Scheffler, M., Hermann, C., Funk, J. M., Wixmerten, A., Rebscher, H., Honig, M., Andreae, C., Buchner, D., Moschel, E., Glathe, A., Jager, E., Thom, M., Greil, A., Bestvater, F., Obermeier, F., Burgmaier, J., Thome, K., Weichert, S., Hein, S., Binnewies, T., Foitzik, V., Muller, M., Stahler, C. F., and Stahler, P. F. (2003). Validation of a novel, fully integrated and flexible microarray benchtop facility for gene expression profiling. *Nucleic Acids Research* 31(23):e151.

16. Beier, M., Baum. M,, Rebscher, H., Mauritz, R., Wixmerten, A., Stahler, C. F., Muller, M., and Stahler, P. F. (2002). Exploring nature's plasticity

---

[2] Available at http://www.ncbi.nlm.nih.gov/BLAST/
[3] Accompanying web page at http://visitor.ics.uci.edu/genex/cybert/

with a flexible probing tool, and finding new ways for its electronic distribution. *Biochem. Soc. Trans.* 30:78–82.

17. Beier, M., and Hoheisel, J. D. (2002). Analysis of DNA-microarrays produced by inverse in situ oligonucleotide synthesis. *J. Biotechnol.* 94:15–22.

18. Ben-Hur, A., Elisseeff, A., and Guyon, I. (2002). A Stability Based Method for Discovering Structure in Clustered Data. *Pacific Symposium on Biocomputing* 2002:6–17.[4]

19. Bender, R., and Lange, S. (2001). Adjusting for multiple testing—when and how? *Journal of Clinical Epidemiology* 54:343–349.

20. Benjamini, Y, and Hochberg, Y. (1995). Controlling the false discovery rate: a practical and powerful approach to multiple testing. *J. R. Statist. Soc. B* 57(1):289–300.

21. Bicciato, S., Pandin, M., Didone, G., and Di Bello, C. (2003). Pattern identification and classification in gene expression data using an autoassociative neural network model. *Biotechnol. Bioeng.* 81(5):594–606.

22. Birnbaum, K., Benfey, P. N., and Shasha, D. E. (2001). Cis element/transcription factor analysis (cis/TF): A method for discovering transcription factor/cis element relationships. *Genome Res.* 11:1567–1573.

23. Black, M. A., and Doerge, R. W. (2002). Calculation of the minimum number of replicate spots required for detection of significant gene expression fold change in microarray experiments. *Bioinformatics* 18(12):1609–16.

24. Bolstad, B. M., Irizarry, R. A., Astrand, M., and Speed, T. P. (2003). A comparison of normalization methods for high density oligonucleotide array data based on variance and bias. *Bioinformatics* 19(2):185–193.

25. Borodovsky, M., and McIninch, J. (1993). GeneMark: Parallel gene recognition for both DNA Strands. *Computers & Chemistry* 17:123–133.

26. David Bowtell and Joseph Sambrook (editors). (2002). *DNA Microarrays: A Molecular Cloning Manual.* New York: Cold Spring Harbor Laboratory Press.

27. Brazma, A., Jonassen, I., Vilo, J., and Ukkonen, E. (1998). Predicting gene regulatory elements in silico on a genomic scale. *Genome Research* 8:1202–1215.

[4] Available online at http://psb.stanford.edu

28. Brenner, S., Johnson, M., Bridgham, J., Golda, G., Lloyd, D. H., Johnson, D., Luo, S., McCurdy, S., Foy, M., Ewan, M., Roth, R., George, D., Eletr, S., Albrecht, G., Vermaas, E., Williams, S. R., Moon, K., Burcham, T., Pallas, M., DuBridge, R. B., Kirchner, J., Fearon, K., Mao, J., and Corcoran, K.. (2000). Gene expression analysis by massively parallel signature sequencing (MPSS) on microbead arrays. *Nature Biotechnology* 18:630–634.

29. Brunak, S., Engelbrecht, J., and Knudsen, S. (1990). Cleaning up gene databases. *Nature* 343:123.

30. Brunak, S., Engelbrecht, J., and Knudsen, S. (1990). Neural network detects errors in the assignment of mRNA splice sites. *Nucleic Acids Res.* 18:4797–4801.

31. Brunak, S., Engelbrecht, J., and Knudsen, S. (1991). Prediction of human mRNA donor and acceptor sites from the DNA sequence. *Journal of Molecular Biology* 220:49–65.

32. Brown, M. P. S., Grundy, W. N., Lin, D., Cristianini, N., Sugnet, C. W., Furey, T. S., Ares, M., and Haussler, D. (2000). Knowledge-based analysis of microarray gene expression data by using support vector machines *Proc. Natl. Acad. Sci. USA* 97:262–267.

33. Burge C., and Karlin, S. (1997). Prediction of complete gene structures in human genomic DNA. *Journal of Molecular Biology* 268:78–94.

34. Bussemaker, H. J., Li, H., and Siggia, E. D. (2000). Building a dictionary for genomes: Identification of presumptive regulatory sites by statistical analysis. *Proc. Natl. Acad. Sci. USA* 97:10096–10100.

35. Chen, T., He, H. L., and Church, G. M. (1999). Modeling gene expression with differential equations *Pacific Symposium on Biocomputing* 4:29–40.[5]

36. Chiang, D. Y., Brown, P. O., and Eisen, M. B. (2001). Visualizing associations between genome sequences and gene expression data using genome-mean expression profiles. *Bioinformatics* 17(Suppl 1):S49–S55.

37. Christiansen, T., Torkington, N., and Wall, L. (1998). *Perl Cookbook*, 1st ed. Sebastopol, CA: O'Reilly & Associates.

38. Claverie, J.-M. (1999). Computational methods for the identification of differential and coordinated gene expression. *Hum. Mol. Genet.* 8:1821–1832.

---

[5] Available online at http://psb.stanford.edu

39. Cole, S. T. et al. (1998). Deciphering the biology of *Mycobacterium tuberculosis* from the complete genome sequence. *Nature* 393:537–544.

40. De Smet, F., Mathys, J., Marchal, K., Thijs, G., De Moor, B., and Moreau, Y. (2002). Adaptive quality-based clustering of gene expression profiles. *Bioinformatics* 18(5):735–746.

41. D'haeseleer, P., Wen, X., Fuhrman, S., and Somogyi, R. (1999). Linear modeling of mRNA expression levels during CNS development and injury *Pacific Symposium on Biocomputing* 4:41–52.[6]

42. Diaz, E., Ge, Y., Yang, Y. H., Loh, K.C., Serafini, T. A., Okazaki, Y., Hayashizaki, Y., Speed, T. P., Ngai, J., and Scheiffele, P. (2002). Molecular analysis of gene expression in the developing pontocerebellar projection system. *Neuron* 36(3):417–434.

43. Dudoit, S., Fridlyand, J., and Speed, T. P. (2000). Comparison of discrimination methods for the classification of tumors using gene expression data. Technical report #576, June 2000.[7]

44. Dudoit, S., Yang, Y., Callow, M. J., and Speed, T. P. (2000). Statistical methods for identifying differentially expressed genes in replicated cDNA microarray experiments. Technical report #578, August 2000.[8]

45. Dysvik, B, and Jonassen, I. (2001). J-Express: Exploring gene expression data using Java. *Bioinformatics* 17:369–370.[9]

46. Efron, B., Storey, J., and Tibshirani, R. (2001). Microarrays, empirical Bayes methods, and false discovery rates. Technical report. Statistics Department, Stanford University.[10]

47. Efron, B., and Tibshirani, R. (2002). Empirical bayes methods and false discovery rates for microarrays. *Genet. Epidemiol.* 23(1):70–86.

48. Fix, E., and Hodges, J. (1951). Discriminatory analysis, nonparametric discrimination: Consistency properties. Technical report, Randolph Field, Texas: USAF School of Aviation Medicine.

49. Fellenberg, K., Hauser, N. C., Brors, B., Neutzner, A., Hoheisel, J. D., and Vingron, M. (2001), Correspondence analysis applied to microarray data. *Proc. Natl. Acad. Sci. USA* 98:10781–10786.

---

[6] Available online at http://psb.stanford.edu
[7] Available at http://www.stat.berkeley.edu/tech-reports/index.html
[8] Available at http://www.stat.berkeley.edu/tech-reports/index.html
[9] Software available at http://www.ii.uib.no/~bjarted/jexpress/
[10] Manuscript available at http://www-stat.stanford.edu/~tibs/research.html

50. Friedman, N., Linial, M., Nachman, I., and Pe'er, D. (2000). Using Bayesian networks to analyze expression data. *Proc. Fourth Annual International Conference on Computational Molecular Biology (RE-COMB)* 2000.

51. Fujibuchi, W., Anderson, J. S. J., and Landsman, D. (2001). PROSPECT improves cis-acting regulatory element prediction by integrating expression profile data with consensus pattern searches. *Nucleic Acids Res.* 29:3988–3996.

52. Gautier, L., Cope, L., Bolstad B.M., and Irizarry R. A. (2003) affy - Analysis of Affymetrix GeneChip data at the probe level. Bioinformatics, in press.

53. Getz, G., Levine, E., and Domany, E. (2000). Coupled two-way clustering analysis of gene microarray data. *Proc. Natl. Acad. Sci. USA* 97:12079–12084.

54. Ghosh, D. (2002). Singular Value Decomposition Regression Models for Classification of Tumors from Microarray Experiments. *Pacific Symposium on Biocomputing* 2002:18–29.[11]

55. Giles, P. J., and Kipling, D. (2003). Normality of oligonucleotide microarray data and implications for parametric statistical analyses. *Bioinformatics* 19:2254–2262 .

56. Goryachev, A. B., Macgregor, P. F., and Edwards, A. M. (2001). Unfolding of microarray data. *Journal of Computational Biology* 8:443–461.

57. Grosu, P., Townsend, J. P., Hartl, D. L., and Cavalieri, D. (2002). Pathway Processor: A tool for integrating whole-genome expression results into metabolic networks. *Genome Research* 12(7):1121–1126.

58. Guigo, R., Knudsen, S., Drake, N., and Smith. T. (1992). Prediction of gene structure. *Journal of Molecular Biology* 226:141–157.

59. Hartemink, A. J., Gifford, D. K., Jaakkola, T. S., and Young, R. A. (2001). Using graphical models and genomic expression data to statistically validate models of genetic regulatory networks. *Pacific Symposium on Biocomputing* 6:422–433.[12]

60. Hartemink, A. J., Gifford, D. K., Jaakkola, T. S., and Young R. A. (2002). Combining Location and Expression Data for Principled Discovery of Genetic Regulatory Network Models. *Pacific Symposium on Biocomputing* 2002:437–449.[13]

---

[11] Available online at http://psb.stanford.edu
[12] Available online at http://psb.stanford.edu
[13] Available online at http://psb.stanford.edu

61. Hacia, J. G. (1999). Resequencing and mutational analysis using oligonucleotide microarrays (Review). *Nat. Genet.* 21(1 Suppl):42–47.

62. Hastie, T., Tibshirani, R., Eisen, M. B., Alizadeh, A., Levy, R., Staudt, L., Chan, W. C., Botstein, D., and Brown, P. (2000). Gene shaving as a method for identifying distinct sets of genes with similar expression patterns. *Genome Biol.* 1:RESEARCH0003.1–21

63. Hastie, T., Tibshirani, R., Botstein, D., and Brown, P. (2001). Supervised harvesting of expression trees. *Genome Biol.* 2:RESEARCH0003.

64. Herrero, J., Valencia, A., and Dopazo, J. (2001). A hierarchical unsupervised growing neural network for clustering gene expression patterns. *Bioinformatics* 17:126–136.

65. von Heydebreck, A., Huber, W., Poustka, A., and Vingron, M. (2001). Identifying splits with clear separation: A new class discovery method for gene expression data. *Bioinformatics* 17(Suppl 1):S107–S114.

66. Holter, N. S., Maritan, A., Cieplak, M., Fedoroff, N. V., and Banavar, J. R. (2000). Dynamic modeling of gene expression data. *Proc. Natl. Acad. Sci. USA* 98:1693–1698.

67. Holter, N. S., Mitra, M., Maritan, A., Cieplak, M., Banavar, J. R., and Fedoroff, N.V. (2000). Fundamental patterns underlying gene expression profiles: Simplicity from complexity. *Proc. Natl. Acad. Sci. USA* 97:8409–8414.

68. Hu, G. K., Madore, S. J., Moldover, B., Jatkoe, T., Balaban, D., Thomas, J., and Wang, Y. (2001). Predicting splice variant from DNA chip expression data. *Genome Research* 11:1237–1245.

69. Huber, W., Von Heydebreck, A., Sultmann, H., Poustka, A., Vingron, M. (2002). Variance stabilization applied to microarray data calibration and to the quantification of differential expression. *Bioinformatics* 18 Suppl 1:S96–S104.

70. Hughes, T. R., Marton, M. J., Jones, A. R., Roberts, C. J., and Stoughton, R., et. al. (2000). Functional discovery via a compendium of expression profiles. *Cell* 102:109–126.

71. Hughes, T. R., Mao, M., Jones, A. R., Burchard, J., Marton, M. J., Shannon, K. W., Lefkowitz, S. M., Ziman, M., Schelter, J. M., Meyer, M. R., Kobayashi, S., Davis, C., Dai, H., He, Y. D., Stephaniants, S. B., Cavet, G., Walker, W. L., West, A., Coffey, E., Shoemaker, D. D., Stoughton, R., Blanchard, A. P., Friend, S. H., and Linsley, P. S. (2001). Expression profiling using microarrays fabricated by an ink-jet oligonucleotide synthesizer. *Nature Biotechnology* 19:342–347.

72. Ideker, T., Thorsson, V., Siegel, A. F., and Hood, L. (2000). Testing for differentially-expressed genes by maximum-likelihood analysis of microarray data. *Journal of Computational Biology* 7:805–817.

73. Ideker, T. E., Thorsson, V., and Karp, R. M. (2000). Discovery of regulatory interactions through perturbation: Inference and experimental design. *Pacific Symposium on Biocomputing* 5:305–316.[14]

74. Ideker, T., Thorsson, V., Ranish, J. A., Christmas, R., Buhler, J., Eng, J. K., Bumgarner, R., Goodlett, D. R., Aebersold, R., and Hood, L. (2001). Integrated genomic and proteomic analyses of a systematically perturbed metabolic network. *Science* 292:929–934.

75. Irizarry, R. A., Bolstad, B. M., Collin, F., Cope, L. M., Hobbs, B., Speed, T. P. (2003). Summaries of Affymetrix GeneChip probe level data. *Nucleic Acids Research* 31(4):e15.

76. Irizarry, R. A., Hobbs, B., Collin, F., Beazer-Barclay, Y. D., Antonellis, K. J., Scherf, U., and Speed, T. P. (2003). Exploration, normalization, and summaries of high density oligonucleotide array probe level data. *Biostatistics* 4(2):249–264.

77. Jarmer, H., Friis, C., Saxild, H. H., Berka, R., Brunak, S., and Knudsen, S. (2002). Inferring parsimonious regulatory networks in *B. subtilis*. *Pacific Symposium on Biocomputing* 2002. Poster presentation.

78. Jenssen, T. K., Laegreid, A., Komorowski, J., and Hovig, E. (2001). A literature network of human genes for high-throughput analysis of gene expression. *Nature Genetics* 28:21–28.

79. Jensen, L. J., and Knudsen, S. (2000). Automatic discovery of regulatory patterns in promoter regions based on whole cell expression data and functional annotation. *Bioinformatics* 16:326–333.

80. Jensen, L. J., Gupta, R., Blom, N., Devos, D., Tamames, J., Kesmir, C., Nielsen, H., Stærfeldt, H. H., Rapacki, K., Workman, C., Andersen, C. A. F., Knudsen, S., Krogh, A., Valencia, A., and Brunak., S. (2002). Ab initio prediction of human orphan protein function from post-translational modifications and localization features. *Journal of Molecular Biology* 319:1257–1265.

81. van Kampen, A. H., van Schaik, B. D., Pauws, E., Michiels, E. M., Ruijter, J. M., Caron, H. N., Versteeg, R., Heisterkamp, S. H., Leunissen, J. A., Baas, F., and van der Mee M. (2000). USAGE: A web-based approach towards the analysis of SAGE data. *Bioinformatics*. 16:899–905.

---

[14]Available online at http://psb.stanford.edu

82. Kanehisa, M., Goto, S., Kawashima, S., and Nakaya, A. (2002). The KEGG databases at GenomeNet. *Nucleic Acids Research* 30(1):42–46. [15]

83. Kerr, M. K., Martin, M., and Churchill, G. A. (2000). Analysis of variance for gene expression microarray data. *J. Comput. Biol.* 7:819–837.

84. Kerr, M. K., and Churchill, G. A. (2001a). Statistical design and the analysis of gene expression microarray data. *Genet Res.* 77:123–128. Review.

85. Kerr, M. K., and Churchill, G. A. (2001b). Bootstrapping cluster analysis: Assessing the reliability of conclusions from microarray experiments. *Proc. Natl. Acad. Sci. USA* 98:8961–8965.

86. Khan, J., Wei, J. S., Ringner, M., Saal, L. H., Ladanyi, M., Westermann, F., Berthold, F., Schwab, M., Antonescu, C. R., Peterson, C., and Meltzer, P. S. (2001). Classification and diagnostic prediction of cancers using gene expression profiling and artificial neural networks. *Nature Genetics* 7:673–679.

87. Kim, S., Dougherty, E. R., Bittner, M. L., Chen, Y., Sivakumar, K., Meltzer, P., and Trent, J. M. (2000). General nonlinear framework for the analysis of gene interaction via multivariate expression arrays. *J. Biomed. Opt.* 5:411–424.

88. Kim, S., Dougherty, E. R., Chen, Y., Sivakumar, K., Meltzer, P., Trent, J. M., and Bittner, M. (2000). Multivariate measurement of gene expression relationships. *Genomics* 15:201–209.

89. Knudsen, S. (1999). Promoter2.0: For the recognition of PolII promoter sequences. *Bioinformatics* 15:356–361. [16]

90. Knudsen, S., Nielsen, H.B., Nielsen, C., Thirstrup, K., Blom, N., Sicheritz-Ponten, T., Gautier, L., Workman, C., and Brunak, S. (2001). T-cell transcriptional responses to HIV infection in vitro. Unpublished.

91. Knudsen, S., Workman, C., Sicheritz-Ponten, T., and Friis, C. (2003). GenePublisher: Automated Analysis of DNA Microarray Data. *Nucleic Acids Research* 31(13):3471–3476. [17]

92. Kohonen, T. (1995). *Self-Organizing Maps*. Berlin: Springer.

---

[15] Available at http://www.genome.ad.jp
[16] Available as a web server at http://www.cbs.dtu.dk/services/Promoter/
[17] Available at http://www.cbs.dtu.dk/services/GenePublisher

93. Krogh, A. (1997). Two methods for improving performance of an HMM and their application for gene finding. *Proc. Fifth Int. Conf. on Intelligent Systems for Molecular Biology (ISMB)* Menlo Park, CA: AAAI Press, pp. 179–186.

94. Krull, M., Voss, N., Choi, C., Pistor, S., Potapov, A., and Wingender, E. (2003). TRANSPATH: an integrated database on signal transduction and a tool for array analysis. *Nucleic Acids Research* 31(1):97–100.

95. Kyoda, K. M., Morohashi, M., Onami, S., and Kitano, H. (2000). A gene network inference method from continuous-value gene expression data of wild-type and mutants. *Genome Informatics* 11:196–204.

96. Lash, A. E., Tolstoshev, C. M., Wagner, L., Schuler, G. D., Strausberg, R. L., Riggins, G. J., and Altschul, S. F. (2000). SAGEmap: A public gene expression resource. *Genome Res.* 10:1051–1060.

97. Lawrence, C. E., Altschul, S. F., Boguski, M. S., Liu, J. S., Neuwald, A. F., and Wootton, J. C. (1993). Detecting subtle sequence signals: A Gibbs sampling strategy for multiple alignment. *Science* 262:208–214.

98. Lazaridis, E. N., Sinibaldi, D., Bloom, G., Mane, S., and Jove, R. (2002). A simple method to improve probe set estimates from oligonucleotide arrays. *Math. Biosci.* 176(1):53–58.

99. Lee, M. L., and Whitmore, G. A. (2002). Power and sample size for DNA microarray studies. *Stat. Med.* 21(23):3543–3570.

100. Lemon, W. J., Palatini, J. T., Krahe, R., and Wright, F. A. (2002). Theoretical and experimental comparisons of gene expression indexes for oligonucleotide arrays. *Bioinformatics* 18(11):1470–1476.

101. Liu, X., Brutlag, D. L., and Liu, J. S. (2001). BioProspector: Discovering conserved DNA motifs in upstream regulatory regions of co-expressed genes. *Pacific Symposium on Biocomputing* 6:127–138.[18]

102. Li, C., and Wong, W. H. (2001a). Model-based analysis of oligonucleotide arrays: Expression index computation and outlier detection. *Proc. Natl. Acad. Sci. USA* 98:31–36.[19]

103. Li, C., and Wong, W. H. (2001b). Model-based analysis of oligonucleotide arrays: Model validation, design issues and standard error application. *Genome Biology* 2:1–11.[20]

[18]Available online at http://psb.stanford.edu
[19]Software available at http://www.dchip.org
[20]Software available at http://www.dchip.org

104. Li, F., and Stormo, G. D. (2001). Selection of optimal DNA oligos for gene expression arrays. *Bioinformatics* 17:1067–1076.

105. Liang, S., Fuhrman S., and Somogyi, R. (1998). REVEAL, A general reverse engineering algorithm for inference of genetic network architectures. *Pacific Symposium on Biocomputing* 3:18–29.[21]

106. Lipshutz, R. J., Fodor, S. P. A., Gingeras, T. R., and Lockhart, D. J. (1999). High density synthetic oligonucleotide arrays. *Nature Genetics Chipping Forecast* 21:20–24.

107. Lockhart, D. J., Dong, H., Byrne, M. C., Follette, M. T., Gallo, M. V., Chee, M. S., Mittmann, M., Wang C., Kobayashi, M., Horton, H., and Brown, E. L. (1996). Expression monitoring by hybridization to high-density oligonucleotide arrays. *Nature Biotechnology* 14:1675–1680.

108. Lonnstedt, I. and Speed, T. (2002). Replicated Microarray Data. *Statistica Sinica* 12:31–46.

109. Lu, P., Nakorchevskiy, A., and Marcotte, E. M. (2003). Expression deconvolution: A reinterpretation of DNA microarray data reveals dynamic changes in cell populations. *Proc. Natl. Acad. Sci. USA* 100(18):10370–10375.

110. Maki, Y., Tominaga, D., Okamoto, M., Watanabe, S., and Eguchi, Y. (2001). Development of a system for the inference of large scale genetic networks. *Pacific Symposium on Biocomputing* 6:446–458.[22]

111. Man, M. Z., Wang, X., and Wang, Y. (2000). POWER_SAGE: Comparing statistical tests for SAGE experiments. *Bioinformatics*. 16:953–959.

112. Margulies, E. H., and Innis, J. W. (2000). eSAGE: Managing and analysing data generated with serial analysis of gene expression (SAGE). *Bioinformatics*. 16:650–651.

113. Marie, R., Jensenius, H., Thaysen, J. Christensen, C. B., and Boisen, A. (2002) . Adsorption kinetics and mechanical properties of thiol-modied DNA-oligos on gold investigated by microcantilever sensors. *Ultramicroscopy* 91:29–36.

114. Masys, D. R., Welsh, J. B., Lynn Fink, J., Gribskov, M., Klacansky, I., and Corbeil, J. (2001). Use of keyword hierarchies to interpret gene expression patterns. *Bioinformatics*. 17:319–326.[23]

[21] Available online at http://psb.stanford.edu
[22] Available online at http://psb.stanford.edu
[23] Web-based software available at http://array.ucsd.edu/hapi/

115. Matthews, B. W. (1975). Comparison of the predicted and observed secondary structure of T4 phage lysozyme. *Biochim. Biophys. Acta* 405:442–451.

116. Matys, V., Fricke, E., Geffers, R., Gossling, E., Haubrock, M., Hehl, R., Hornischer, K., Karas, D., Kel, A. E., Kel-Margoulis, O. V., Kloos, D. U., Land, S., Lewicki-Potapov, B., Michael, H., Munch, R., Reuter, I., Rotert, S., Saxel, H., Scheer, M., Thiele, S., and Wingender, E. (2003). TRANSFAC: Transcriptional regulation, from patterns to profiles. *Nucleic Acids Research* 31(1):374–378.

117. Matz, M., Usman, N., Shagin, D., Bogdanova, E., and Lukyanov, S. (1997). Ordered differential display: A simple method for systematic comparison of gene expression profiles. *Nucleic Acids Res.* 25:2541–2542.

118. McKendry, R., Zhang, J., Arntz, Y., Strunz, T., Hegner, M., Lang, H. P., Baller, M. K., Certa, U., Meyer, E., Guntherodt, H. J., and Gerber, C. (2002). Multiple label-free biodetection and quantitative DNA-binding assays on a nanomechanical cantilever array. *Proc. Natl. Acad. Sci. USA* 99:9783–9788.

119. Michaels, G. S., Carr, D. B., Askenazi, M., Fuhrman, S., Wen, X., and Somogyi, R. (1998). Cluster analysis and data visualization of large-scale gene expression data. *Pacific Symposium on Biocomputing* 3:42–53.[24]

120. Montgomery, D. C., and Runger, G. C. (1999). *Applied Statistics and Probability for Engineers.* New York: Wiley.

121. Neuwald, A. F., Liu, J. S., and Lawrence, C. E. (1995). Gibbs motif sampling: Detection of bacterial outer membrane protein repeats. *Protein Science* 4:1618–1632.

122. Newton, M. A., Kendziorski, C. M., Richmond, C. S., Blattner, F. R., and Tsui, K. W. (2001). On differential variability of expression ratios: Improving statistical inference about gene expression changes from microarray data. *J. Comput. Biol.* 8:37–52.

123. Nielsen, H. B., and Knudsen, S. (2002). Avoiding cross hybridization by choosing nonredundant targets on cDNA arrays. *Bioinformatics* 18:321–322.

124. Nielsen, H. B., Wernersson, R., and Knudsen, S. (2003). Design of oligonucleotides for microarrays and perspectives for design of multi-transcriptome arrays. *Nucleic Acids Research* 31:3491–3496.

[24] Available online at http://psb.stanford.edu

125. Noordewier, M. O., and Warren, P. V. (2001). Gene expression microarrays and the integration of biological knowledge. *Trends. Biotechnol.* 19:412–415.

126. Nuwaysir, E. F., Huang, W., Albert, T. J., Singh, J., Nuwaysir, K., Pitas, A., Richmond, T., Gorski, T., Berg, J. P., Ballin, J., McCormick, M., Norton, J., Pollock, T., Sumwalt, T., Butcher, L., Porter, D., Molla, M., Hall, C., Blattner, F., Sussman, M. R., Wallace, R. L., Cerrina, F., and Green, R. D. (2002). Gene expression analysis using oligonucleotide arrays produced by maskless photolithography. *Genome Research* 12:1749–1755.

127. Ouzounis, C. A., and Valencia, A. (2003). Early bioinformatics: The birth of a discipline—a personal view. *Bioinformatics* 19(17):2176–2190.

128. Pan, W., Lin, J., and Le, C. (2001). How many replicates of arrays are required to detect gene expression changes in microarray experiments? A mixture model approach. Report 2001-012, Division of Biostatistics, University of Minnesota.[25]

129. Pan, W., Lin, J., and Le, C. (2002). How many replicates of arrays are required to detect gene expression changes in microarray experiments? A mixture model approach. *Genome Biology* 3(5):research0022.

130. Parmigiani, G., Garrett, E. S., Irizarry, R. A., and Zeger, S. (Editors), (2003) *The Analysis of Gene Expression Data* Berlin: Springer Verlag.

131. Park, P. J., Pagano, M., and Bonetti, M. (2001). A nonparametric scoring algorithm for identifying informative genes from microarray Data. *Pacific Symposium on Biocomputing* 6:52–63. [26]

132. Pavlidis, P., Li, Q., and Noble, W. S. (2003). The effect of replication on gene expression microarray experiments. *Bioinformatics* 19(13):1620–1627.

133. Pearson, W. R. (2001). Training for Bioinformatics and Computational Biology (Editorial). *Bioinformatics* 17:761–762.

134. Pe'er, D., Regev, A., Elidan, G., and Friedman, N. (2001). Inferring subnetworks from perturbed expression profiles. *Bioinformatics* 17 (Suppl. 1):S215–S224.

135. Periwal, V., and Szallasi, Z. (2002). Trading "wet-work" for network. Nature Biotechnology 20:345–346.

---

[25] Available at http://www.biostat.umn.edu/cgi-bin/rrs?print+2001
[26] Manuscript available online at http://psb.stanford.edu

136. Piper, M. D., Daran-Lapujade, P., Bro, C., Regenberg, B., Knudsen, S., Nielsen, J., Pronk, J. T. (2002). Reproducibility of oligonucleotide microarray transcriptome analyses. An interlaboratory comparison using chemostat cultures of Saccharomyces cerevisiae. *J. Biol. Chem.* 277(40):37001–37008.

137. Rain, J. C., Selig, L., De Reuse, H., Battaglia, V., Reverdy, C., Simon, S., Lenzen, G., Petel, F., Wojcik, J., Schachter, V., Chemama, Y., Labigne, A., and Legrain, P. (2001). The protein-protein interaction map of *Helicobacter pylori*. *Nature* 409:211–215.

138. Raychaudhuri, S., Stuart, J. M., and Altman, R. B. (2000). Principal components analysis to summarize microarray experiments: Application to sporulation time series. *Pac. Symp. Biocomput.* 2000:455–466.[27]

139. Reiner, A., Yekutieli, D., Benjamini, Y. (2003). Identifying differentially expressed genes using false discovery rate controlling procedures. *Bioinformatics* 19(3):368–375.

140. Roberts, C. J., Nelson, B., Marton, M. J., Stoughton, R., Meyer, M. R., Bennett, H. A., He, Y. D. D. , Dai, H. Y., Walker, W. L., Hughes, T. R., Tyers, M., Boone, C., and Friend, S. H. (2000). Signaling and circuitry of multiple MAPK pathways revealed by a matrix of global gene expression profiles. *Science* 287:873–880.

141. Rocke, D. M., and Durbin, B. (2001). A model for measurement error for gene expression arrays. *Journal of Computational Biology* 8(6):557–569.

142. Rouillard, J. M., Herbert, C. J., and Zuker, M. (2002). OligoArray: genome-scale oligonucleotide design for microarrays. *Bioinformatics* 18(3):486–487.

143. Samsonova, M. G., and Serov, V. N. (1999). NetWork: An interactive interface to the tools for analysis of genetic network structure and dynamics. *Pacific Symposium on Biocomputing* 4:102–111.[28]

144. Sasik, R., Hwa, T., Iranfar, N., and Loomis, W. F. (2001). Percolation clustering: A novel algorithm applied to the clustering of gene expression patterns in dictyostelium development. *Pacific Symposium on Biocomputing* 6:335–347.[29]

145. Segal, E., Taskar, B., Gasch, A., Friedman, N., and Koller, D. (2001). Rich probabilistic models for gene expression. *Bioinformatics* 17(Suppl 1):S243–S252.

---

[27] Available online at http://psb.stanford.edu
[28] Available online at http://psb.stanford.edu
[29] Available online at http://psb.stanford.edu

146. Schadt, E. E., Li, C., Su, C., and Wong, W. H. (2000). Analyzing high-density oligonucleotide gene expression array data. *J. Cell. BioChem.* 80:192–201.

147. Schena, Mark. (1999). *DNA microarrays: A practical approach* (Practical Approach Series, 205). Oxford: Oxford Univ. Press.

148. Schena, Mark. (2000). *Microarray Biochip Technology.* Sunnyvale, CA: Eaton.

149. Scherf, M., Klingenhoff, A., and Werner, T. (2000). Highly specific localization of promoter regions in large genomic sequences by PromoterInspector: A novel context analysis approach. *Journal of Molecular Biology* 297:599–606.

150. Schuchhardt, J., Beule, D., Malik, A., Wolski, E., Eickhoff, H., Lehrach, H., and Herzel, H. (2000). Normalization strategies for cDNA microarrays. *Nucleic Acids Res.* 28:E47.

151. Schwartz, R. L. and Phoenix, T. (2001). *Learning Perl*, 3rd ed. Sebastopol, CA: O'Reilly & Associates

152. Shannon, M. F., and Rao S. (2002). Transcription. Of chips and ChIPs. *Science* 296(5568):666–669.

153. Sheng, Q., Moreau, Y., and De Moor, B. (2003). Biclustering microarray data by Gibbs sampling. *Bioinformatics* 19(Suppl 2):II196–II205.

154. Shrager, J., Langley, P., and Pohorille, A. (2002). Guiding Revision of Regulatory Models with Expression Data *Pacific Symposium on Biocomputing* 2002:486–497.[30]

155. Singh-Gasson, S., Green, R. D., Yue, Y., Nelson, C., Blattner, F., Sussman, M. R., and Cerrina F. (1999). Maskless fabrication of light-directed oligonucleotide microarrays using a digital micromirror array. *Nature Biotechnology* 17:974–978.

156. Skovgaard, M., Jensen, L. J., Brunak, S., Ussery, D., and Krogh, A. (2001). On the total number of genes and their length distribution in complete microbial genomes. *Trends Genet.* 17:425–428.

157. van Someren, E. P., Wessels, L. F. A., and Reinders, M. J. T. (2000). Linear modeling of genetic networks from experimental data. *Proc. ISMB* 2000:355–366.

---

[30] Available online at http://psb.stanford.edu

158. Speed, T. P., and Yang, Y. H. (2002) Direct versus indirect designs for cDNA microarray experiments Technical report #616, Department of Statistics, University of California, Berkeley.

159. Spellman, P., Sherlock, G., Zhang, M., Lyer, V., Anders, K., Eisen, M., Brown, P., Botstein, D., and Futcher, B. (1998). Comprehensive identification of cell cycle-regulated genes of yeast *S. cerevisiae* by microarray hybridization. *Mol. Biol. Cell* 9:3273–3297.

160. Spicker, J. S., Wikman, F, Lu M. L., Cordon-Cardo C., Ørntoft, T. F., Brunak, S., and Knudsen, S. (2002). Neural network predicts sequence of TP53 gene based on DNA chip. *Bioinformatics* 18:1133–1134.

161. Szallasi, Z. (1999). Genetic network analysis in light of massively parallel biological data acquisition. *Pacific Symposium on Biocomputing* 4:5–16.[31]

162. Tamayo, P., Slonim, D., Mesirov, J., Zhu, Q., Kitareewan, S., Dmitrovsky, E., Lander, E. S., and Golub, T. R. (1999). Interpreting patterns of gene expression with self-organizing maps: Methods and application to hematopoietic differentiation. *Proc. Natl. Acad. Sci. USA* 96:2907–2912.

163. Tanabe, L., Scherf, U., Smith, L. H., Lee, J. K., Hunter, L., and Weinstein, J. N. (1999). MedMiner: An internet text-mining tool for biomedical information, with application to gene expression profiling. *BioTechniques* 27:1210–1217.[32]

164. Tanay, A., and Shamir, R. (2001). Expansion on existing biological knowledge of the network: Computational expansion of genetic networks. *Bioinformatics* 17(Suppl 1):S270–S278.

165. Theilhaber, J., Bushnell, S., Jackson, A., Fuchs, R. (2001). Bayesian estimation of fold-changes in the analysis of gene expression: the PFOLD algorithm. *Journal of Computational Biology* 8(6):585–614.

166. Thieffry, D., and Thomas, R. (1998). Qualitative analysis of gene networks. *Pacific Symposium on Biocomputing* 3:77–88.[33]

167. Tibshirani, R., Walther, G., Botstein, D., and Brown, P. (2000). Cluster validation by prediction strength. Technical report. Statistics Department, Stanford University.[34]

---

[31]Available online at http://psb.stanford.edu
[32]Web version available at http://discover.nci.nih.gov/textmining/filters.html
[33]Available online at http://psb.stanford.edu
[34]Manuscript available at http://www-stat.stanford.edu/~tibs/research.html

168. Thomas, J. G., Olson, J. M., Tapscott, S. J., and Zhao, L. P. (2001). An efficient and robust statistical modeling approach to discover differentially expressed genes using genomic expression profiles. *Genome Res.* 11:1227–1236.

169. Thykjaer, T., Workman, C., Kruhøffer, M., Demtröder, K., Wolf, H., Andersen, L. D., Frederiksen, C. M., Knudsen, S., and Ørntoft, T. F. (2001). Identification of gene expression patterns in superficial and invasive human bladder cancer. *Cancer Research* 61:2492–2499.

170. Tusher, V. G., Tibshirani, R., and Chu, G. (2001). Significance analysis of microarrays applied to the ionizing radiation response. *Proc. Natl. Acad. Sci. USA* 98:5119–5121.[35]

171. Varotto, C., Richly, E., Salamini, F., and Leister, D. (2001). GST-PRIME: A genome-wide primer design software for the generation of gene sequence tags. *Nucleic Acids Res.* 29:4373–4377.

172. van't Veer, L. J., Dai, H., van de Vijver, M. J., He, Y. D., Hart, A. A., Mao, M., Peterse, H. L., van der Kooy, K., Marton, M. J., Witteveen, A. T., Schreiber, G. J., Kerkhoven, R. M., Roberts, C., Linsley, P. S., Bernards, R., Friend, S. H. (2002). Gene expression profiling predicts clinical outcome of breast cancer. *Nature* 415:530–536.

173. Velculescu, V. E., Zhang, L., Vogelstein, B., and Kinzler, K. W. (1995). Serial analysis of gene expression. *Science.* 270:484–487.

174. Venables, W. N. and Ripley, B. D. (1999). *Modern applied statistics with S-PLUS*, 3rd ed. New York, NY: Springer.

175. Venet, D., Pecasse, F., Maenhaut, C., and Bersini, H. (2001). Separation of samples into their constituents using gene expression data. *Bioinformatics* 17(Suppl 1):S279–287.

176. Vingron, M. (2001). Bioinformatics Needs to adopt statistical thinking (Editorial). *Bioinformatics* 17:389–390.

177. Wagner, A. (2001). How to reconstruct a large genetic network from n gene perturbations in fewer than n(2) easy steps. *Bioinformatics* 17(12):1183–1197.

178. Wagner, A. (2002). Estimating coarse gene network structure from large-scale gene perturbation data. *Genome Research* 12(2):309–315.

179. Wahde, M., and Hertz, J. (2000). Coarse-grained reverse engineering of genetic regulatory networks. *Biosystems* 55:129–136.

[35] Software available for download at http://www-stat.stanford.edu/~tibs/SAM/index.html

180. Wahde, M., and Hertz, J. (2001). Modeling genetic regulatory dynamics in neural development. *J. Comput. Biol.* 8:429–442.

181. Wahde, M., Klus, G. T., Bittner, M. L., Chen, Y., Szallasi, Z. (2002). Assessing the significance of consistently mis-regulated genes in cancer associated gene expression matrices. *Bioinformatics* 18(3):389-394.

182. Wall, L., Christiansen, T., and Orwant, J. (2000). Programming Perl, 3rd ed. Sebastopol, CA: O'Reilly & Associates.

183. Wall, M. E., Dyck, P. A., and Brettin, T. S. (2001). SVDMAN—singular value decomposition analysis of microarray data. *Bioinformatics* 17:566–568.

184. Weaver, D., Workman, C., and Stormo, G. (1999). Modeling regulatory networks with weight matrices. *Pacific Symposium on Biocomputing* 4:122–123.[36]

185. Wessels, L. F. A., Van Someren, E. P., and Reinders, M. J. T. (2001) A Comparison of genetic network models. *Pacific Symposium on Biocomputing* 6:508–519.[37]

186. Wikman, F. P., Lu, M. L., Thykjaer, T., Olesen, S. H., Andersen, L. D., Cordon-Cardo, C., and Ørntoft, T. F. (2000). Evaluation of the performance of a p53 sequencing microarray chip using 140 previously sequenced bladder tumor samples. *Clin. Chem.* 46:1555–1561.

187. Wodicka, L., Dong, H., Mittmann, M., Ho, M. H., and Lockhart, D. J. (1997). Genome-wide expression monitoring in Saccharomyces cerevisiae. *Nature Biotechnology* 15:1359–1367.

188. Wolfinger, R. D., Gibson, G., Wolfinger, E. D., Bennett, L., Hamadeh, H., Bushel, P., Afshari, C., Paules, R. S. (2001). Assessing gene significance from cDNA microarray expression data via mixed models. *Journal of Computational Biology* 8(6):625–637.

189. Wolfsberg, T. G., Gabrielian, A. E., Campbell, M. J., Cho, R. J., Spouge, J. L., and Landsman, D. (1999). Candidate regulatory sequence elements for cell cycle-dependent transcription in *Saccharomyces cerevisiae*. *Genome Res.* 9:775–792.

190. Workman, C., and Stormo, G.D. (2000) ANN-Spec: A method for discovering transcription factor binding sites with improved specificity. *Pacific Symposium on Biocomputing 2000.*[38]

---

[36] Available online at http://psb.stanford.edu
[37] Available online at http://psb.stanford.edu
[38] Available online at http://psb.stanford.edu

191. Workman, C., Jensen, L.J., Jarmer, H., Berka, R., Saxild, H.H., Gautier, L., Nielsen, C., Nielsen, H.B., Brunak, S, and Knudsen, S. (2002) A new non-linear normalization method for reducing variance between DNA microarray experiments. Genome Biology 3(9):0048.[39]

192. Xia, X, and Xie, Z. (2001). AMADA: Analysis of microarray data. *Bioinformatics* 17:569–570.

193. Xing, E. P., and Karp, R. M. (2001). CLIFF: Clustering of high-dimensional microarray data via iterative feature filtering using normalized cuts. *Bioinformatics* 17(Suppl 1):S306–S315.

194. Xiong, M., Jin, L., Li, W., and Boerwinkle, E. (2000). Computational methods for gene expression-based tumor classification. *Biotechniques* 29:1264–1268.

195. Yamamoto, M., Wakatsuki, T., Hada, A., and Ryo, A. (2001). Use of serial analysis of gene expression (SAGE) technology. *J Immunol. Methods*. 250:45–66. Review.

196. Yang, Y. H., Buckley, M. J., and Speed, T. P. (2001). Analysis of cDNA microarray images. *Briefings in Bioinformatics* 2(4):341–349.

197. Yang, Y. H., Buckley, M. J., Dudoit, S., and Speed, T. P. (2001). Comparison of methods for image analysis on cDNA microarray data. Technical report #584, Department of Statistics, University of California, Berkeley.

198. Yang, Y. H., Dudoit, S., Luu, P., Lin, D. M., Peng, V., Ngai, J., and Speed, T. P. (2002). Normalization for cDNA microarray data: a robust composite method addressing single and multiple slide systematic variation. *Nucleic Acids Research* 30(4):e15.

199. Yeang, C. H., Ramaswamy, S., Tamayo, P., Mukherjee, S., Rifkin, R. M., Angelo, M., Reich, M., Lander, E., Mesirov, J., and Golub, T. (2001). Molecular classification of multiple tumor types. *Bioinformatics* 17(Suppl 1):S316–S322.

200. Yeung, K. Y., Fraley, C., Murua, A., Raftery, A. E., and Ruzzo, W. L. (2001) Model-based clustering and data transformations for gene expression data. *Bioinformatics* 17:977–987.

201. Yeung, K. Y., Haynor, D. R., and Ruzzo, W. L. (2001). Validating clustering for gene expression data. *Bioinformatics* 17:309–318.

---

[39]Software available in affy package of Bioconductor http://www.bioconductor.org

202. Yoo, C., Thorsson, V., and Cooper, G.F. (2002). Discovery of Causal Relationships in a Gene-Regulation Pathway from a Mixture of Experimental and Observational DNA Microarray Data. *Pacific Symposium on Biocomputing* 2002:498–509.[40]

203. Zien, A., Aigner, T., Zimmer, R., and Lengauer, T. (2001). Centralization: A new method for the normalization of gene expression data. *Bioinformatics* 17(Suppl 1):S323–S331.

204. Zhang, L., Miles, M. F., and Aldape, K. D. (2003a). A model of molecular interactions on short oligonucleotide microarrays. *Nature Biotechnology* 21(7):818–821.

205. Zhang, L., Miles, M. F., and Aldape, K. D. (2003b) Corrigendum: A model of molecular interactions on short oligonucleotide microarrays. *Nature Biotechnology* 21(8):941.

206. Zhao, L. P., Prentice, R., and Breeden, L. (2001). Statistical modeling of large microarray data sets to identify stimulus-response profiles. *Proc. Natl. Acad. Sci. USA* 98:5631–5636.

207. Zhu, J., and Zhang, M. Q. (2000). Cluster, function and promoter: Analysis of yeast expression array. *Pacific Symposium on Biocomputing* 5:476–487.[41]

---

[40] Available online at http://psb.stanford.edu
[41] Available online at http://psb.stanford.edu

Index